41313

THE HOME UNIVERSITY LIBRARY
OF MODERN KNOWLEDGE

106

POLITICAL THOUGHT
IN ENGLAND

*The Utilitarians from
Bentham to Mill*

POLITICAL THOUGHT
IN ENGLAND

The Utilitarians
from Bentham to Mill

WILLIAM L. DAVIDSON

Geoffrey Cumberlege
OXFORD UNIVERSITY PRESS
LONDON   NEW YORK   TORONTO

First published in 1915 *and reprinted in* 1929, 1931, 1933, 1935,
1942 *and* 1944
This reset impression published in 1947

320.942

PRINTED IN GREAT BRITAIN

CONTENTS

CHAPTER I

THE UTILITARIAN POSITION

NOT infrequently we find writers, men of letters and philosophers alike, referring to Utilitarianism as they might to Epicureanism of old, as a rounded and completed thing delivered once for all by a master, and handed down full-formed from the beginning, with little or no modification by succeeding generations. But Utilitarianism, like most other philosophical systems, is a growth, beginning with a clear but restricted view, and needing experience and the critical sympathetic insight of others later on to widen its outlook, to tone down its dogmatism, to lop off excrescences, and to adjust it to fresh light and new situations. There is a school or succession of utilitarians, in the same sense as there is a school or succession of intuitional moralists, each with his own merits and peculiarities, and each advancing beyond the other, enriching the common teaching in some respects and carrying it forward to a fuller issue. There is a marked and significant difference, for instance, between the teaching of Bentham and that of J. S. Mill—a difference that hardly finds a parallel in the teaching of Epicurus and the brilliant reproduction of it by his disciple Lucretius.

But, this granted, it must be granted also that there is a point of view common to utilitarians and a common spirit that animates them, and that there are principles generally acknowledged. And so, it may be well to begin with an indication of the utilitarian standpoint, and with a sketch of the leading points of general, if not absolutely universal, agreement.

I. The term Utilitarianism, designative of a philosophical theory in ethics and in politics, is a very modern one; but the thing that it represents is very old. It represents interest in the welfare of mankind, wedded to practical efforts to ameliorate the conditions of human life on rational principles, and to raise the masses through effective State legislation. Not all utilitarians have been men of emotion: intellect has been more conspicuous than sentiment in many of the leaders. But all have had at heart the general welfare, and have aimed as best they could at promoting it. The name found currency through J. S. Mill, and has been in constant use since his time. He does not claim to have invented it, but to have adopted it from a passing expression in Galt's *Annals of the Parish* (see note to Chapter II of his *Utilitarianism*). But he need not have gone to Galt for it: he might have found it in Bentham, who uses it twice in his writings. Anyhow, the name, when explicitly adopted by Mill and his fellow-thinkers, was bound to give offence to philosophy in its strict acceptation, because the last thing that the philosopher as pure thinker would occupy himself with is human welfare in general and practical reform. The philosopher, so Plato had taught, is a veritable 'innocent abroad' when brought into relation with the busy world and the practical affairs of life, and draws only ridicule upon himself when he has to play his part as an active citizen. The world in which *he* lives is that of abstract speculation and contemplation, not the concrete world of action and every-day concerns. He has neither knowledge of, nor interest in, what is mundane. He is unacquainted with his next-door neighbour, and is wholly ignorant of the social events that are passing around him. Sometimes, indeed, he is compelled to take part in common life, and then he makes himself ludicrous. 'On every occasion, private as well as public . . . when he appears in a law-court, or in any place in which he has to speak of things which are at his feet, and before his eyes, he is the jest not only of Thracian handmaids [as Thales was] but of the general herd, tumbling into wells and every sort of disaster through his inexperience' (Theaetetus).

Almost the exact opposite of all this characterizes the utilitarian. The truths of intellect, indeed, are his concern; but the needs and interests of life are regarded by him as first and supreme: his theme is the happiness of men and how it may be effected. He does not stand aloof from the active toiling world, but is at home in it. The concrete, not the abstract, is what most attracts him; and man as a 'social' being takes precedence in his estimation of man as a solitary thinker. He is by temperament and by conviction a pragmatist—practical and concrete; valuing ideas mainly in so far as they work, and in so far as they serve such purposes as men desire and for which they strive. The meaning of this is that his first and great concern is human life, human activity, human well-being; and, politically, he is the strenuous opponent of tyranny and injustice, and the champion of individual freedom. Hence, utilitarianism is emphatically practical, and keeps in close touch with experience; and it is reformatory, having in view the constant elevation of human life and the furthering of human progress.

But not only were the terms 'utility' and 'utilitarianism' obnoxious to the pure philosopher, they are also unfortunate in themselves, as tending to bring along with them all the associations that cluster round them in the every-day usage of the plain man, and, consequently, to produce confusion and misconception. When employed in connexion with human aspirations and efforts, they are apt to savour of the selfish and the commercial; and thus they suffer from the fact that a certain sordidness attaches in popular estimation to the things that they signify. Utility means serving a purpose; and what is useful often serves a very mean or humble purpose, and its ministry, even though necessary to our comfort, is apt to be despised. It is not easy for us to clothe with the august attribute of *worth* the simple door-mat as we find it in common use, the coal-scuttle, or the dust-bin. Our very dependence, day after day, on these and similar things and their commonplace or non-ideal character tend to make us lightly esteem them. But utility need not be thus contemptuously treated (even the door-mat and the coal-scuttle and the dust-bin have their value), and it need not be thus narrowly restricted in its signification. To every part of man's nature there are corresponding utilities; and a man is to be conceived, not only as an individual, but as an individual who is by nature social—whose very existence and whose continued welfare depend on the existence and co-operation of others, to whom he is linked by bonds of altruism and human affection, and whose claims and interests his own egoism is bound to respect. Moreover, he is a being for whom 'bread and butter' is not the sole end of endeavour, to whom the 'muck rake' is not the only thing that makes appeal: he is stirred by ideals—intellectual, educational, political, ethical, social. And so, utility for him means what is best for all the elements of his nature, and what can most effectively promote his full and ultimate good, and the full and ultimate good of his fellows. Utilitarians have expressed this by saying that utility means 'happiness', or, more completely (after Bentham), 'the greatest happiness of the greatest number', or, again, 'enlightened benevolence'.

In like manner, popular usage has degraded the meaning of the term 'utilitarianism'. It has narrowed it to the lower sphere of human desire and activity, and has weighted it with disparaging associations. Hence, the popular orator, worked into a frenzy, when he wishes to condemn the present age as censoriously as he can, labels it 'a *utilitarian* age'. This he sometimes varies with the phrase, scornfully pronounced, 'a *materialistic* age'; thereby identifying utilitarianism with materialism in its worst ethical sense, as the inordinate and irrational pursuit of wealth and worldly prosperity. One wonders how far Thomas Carlyle, with his vehement rhetoric, is responsible for this!

In order to avoid the misconceptions arising from this popular usage, some have proposed to supplant the terms 'utility' and 'happiness' by such terms as 'welfare' and 'well-being'. These are supposed to be more appropriate and less misleading. And so, indeed, they are. There is a certain natural attractiveness about a man's well-being and welfare that does not attach to his effort to further his own interest or gratify himself, which utility and happiness are not unlikely to suggest.

Moreover, well-being and welfare, besides having in ordinary usage a nobler connotation, are susceptible of a wider application. They imply all that utility implies, and more: they touch the imagination in a way that utility fails to do, and they are not associated with, or do not readily point to, selfish regard to one's own prosperity (which seems implied in happiness), as distinguished from, and in part opposed to, the good and prosperity of others. *My* well-being, even though it is mine, is not selfish if it can be effected only in unison with the well-being of others; and *my* life cannot be ignoble, even if I myself must necessarily be the centre of it, if its expansion is dependent on my regard for and interest in the lives of my fellow men. The individual must stand first to himself, from the very nature of the case; but he cannot be reprehended for this if his development depends on the identification of himself with the aims and ideals of those with whom he associates, or of mankind in general.

Utility, then, is welfare; and welfare covers every conceivable element that goes to determine and constitute man's happiness.

II. In defining further the conception of Utilitarianism, we must consider its relation to Psychology—more especially, the psychology of our moral nature.

In so far as psychology in general is concerned, utilitarianism at the beginning accepted the English tradition, going back to Locke. Its method is inductive, its basis experiential, and its end practical. It refuses to regard man merely as pure intellect, but insists on taking account of his complexity of nature and diversity of interests, recognizing that his interests are determined by his likings and his aspirations. A man's likings are, in the first place, personal—what his own feelings and desires prompt to, on the principle of self-preservation; but, as the individual is from birth a member of society, these feelings and desires have necessary reference to other men and other sentient beings, and can neither exist nor be satisfied apart from them. No greater mistake can be made than that of regarding an individual as simply an individual. A man who is an individual solely, i.e. an absolutely isolated being, is a mere fiction of the mind: there is nothing corresponding to him in reality. Every human being is necessarily the product of parents, who in turn are the children of parents, and these of others, and so on backwards in an endless line; and every individual, helpless for years, is dependent for his early upbringing, for his continued life and for his education (mental, physical, and moral) on others, and he cannot, if he would, discard the social influence and early social environment. His later life, as much as his earlier days, is dependent on human co-operation and on contact with his fellow men; and, if he moulds, he is also moulded by, his human surroundings.

Hence, the question comes to be, What is man, this social being, moved by? What does he aim at? The usual utilitarian answer is, Happiness. But this happiness is not a man's own alone. His own 'good' is realized in conjunction with that of others; and he and they alike are eager for a life of satisfaction and contentment (so it is maintained, especially by Bentham). But a satisfied, contented life is, in the ultimate analysis, a life of 'pleasure'; and so, to the utilitarian, Pleasure is the individual's ultimate end—the aim of his being and the object of his desire.

What this fully means will be seen as we proceed; meanwhile, we note the fact.

But happiness, even a man's own, if, in the circumstances of life, it cannot be obtained independently of regard to others, is necessarily dependent on the existence and organization of the State. Given the position that happiness is universally desired, the attainment of it in a community is conditioned by the encouragement and limitations imposed by custom, law, and legislation. Hence, the utilitarian cannot dissociate the ultimate end of desire from political and State action. Such action both furthers definite duty by giving a distinct stimulus to the discharge of it, and also sanctions it by supplying an authoritative rule and approbation.

In this way, politics to the utilitarian implicates ethics: with him, ethical and political philosophy go together. A political sanction has value only if it has in view the good of those for whom the legislation exists. The welfare of people in general is the supreme consideration; and that implies the removal of hindrances towards the improvement or betterment of the citizens, and also the provision of conditions best suited for the promotion of this betterment. There are both a negative and a positive aspect of proper legislation, viz. the getting rid of degrading or untoward circumstances, and the putting of favourable inducements in their place.

In order to accomplish this, there is, obviously, needed an adequate knowledge of human nature, and, therefore, of the motives by which human beings are swayed and of the ideals that they are tending to realize.

This, in turn, points to the necessity of rendering ethics (the study of human character and conduct) really scientific. Haphazard statements about morality and social life, superficial analyses, wild flights of the imagination, and mere unverified assumptions must be discarded, and serious and systematic investigation instituted concerning the moral side of man's being, and a strenuous attempt essayed to subject ethical phenomena and processes to the conditions that science imposes on *its* investigation, viz. observation and experiment (the latter necessarily limited in ethics and society) and a full and accurate application of the inductive method. Ethical *theory* is made to wait on ethical *fact*, and generalization proceeds only from data afforded by experience. Utilitarian ethics is of necessity analytic, descriptive, and inductive, resting on ascertained facts; and its aim has reference to the right use of the facts, so as to advance social progress and for the concrete purpose of improving the existing conditions of life. An ideal ethics that has nothing to support it in experience is, consequently, disowned—is even contemptuously cast aside; but ideality that aims at practical reform is not only not disowned but supplies the very motive of the utilitarian's efforts to bring about a social millennium, in which both the individual and the race shall find their highest happiness and good.

And here we have the answer to the objection frequently brought against utilitarianism, that it is lacking in ideality—that, inasmuch as it concerns itself with men's actual needs and pressing circumstances, it fails to fire the imagination, and thereby to make up for present discomfort, hardships and sufferings by pictures of a glorious future, when the golden age shall have been reached and pain and misery shall be no more. Man, it is urged by the objector, is moved by ideals, and not by facts: even when an imagined new world is felt to be nothing more than a fancy, the entertaining of it is good for the individual and has worth. But the reply is, that it is a mistake to suppose that the utilitarian works without ideals: on the contrary, the vision of the future improvement of society and of the regeneration of mankind is precisely what inspires and stimulates him and upholds him in the face of difficulties and seeming failure. Only, the ideals that he cherishes are of an essentially practical and human kind. He believes that they are both desirable and capable of realization—they are not solely 'in the clouds'. The ideals that he discards are simply what appear to him to be either undesirable or unrealizable or both. 'A new heaven and a new earth' rises before his mind, as before the mind of the sentimentalist, but it is conditioned by his knowledge of man's constitution—of its actual character, its wants, its possibilities, and its obvious limitations; and he refuses to become either a fanatic or a dreamer. Perhaps, he carries this too far; for sentiment after all has its place, and it may be better for us to cherish a purpose, even though it be an unrealizable purpose, in the 'heart' than not to have planned and purposed at all. But he is clearly justified in exercising the imagination with due appreciation of life's conditions and possibilities rather than in allowing his mind to run riot in devising impossible schemes and dreaming futile unsubstantial dreams. Earth is to him *terra firma*; and any effort at improving man's position here is to be taken in connexion with this fact, and to be shaped in accordance therewith.

This practical character was stamped upon utilitarianism at the very outset. Bentham defined Utility by opposing it to two things, viz. to Asceticism, on the one hand, and, on the other hand, to Sympathy and Antipathy; thereby bringing out both its practical nature and the need of constantly guarding against our being perverted by prejudice or overcome by the emotions. In other words, he wanted utilitarianism to be a working creed, opposed both to the 'cloister'd vertue', with its unnatural hugging of pain, and to the dangerous character of non-rational approbation or dislike.

Utilitarianism, we have just said, is experiential: it founds on experience, and it appeals to experience as the ultimate test. Now, what kind of experience is it to which the utilitarian appeals? In the first place, it is experience as opposed to abstract theory or speculation —to theory divorced from actual trial in life. Theory of that stamp does not pay regard to consequences; and consequences are everything to the utilitarian. To the 'dialectician' of Plato, or to the pure mathematician of the present day, it does not much matter whether an idea *works* or not—perhaps, it is all the better that it does not work; but, to the utilitarian, practical application is an indispensable consideration. In the next place, it is experience regarded as the source and origin of knowledge. Locke had analysed this into 'sensation' and 'reflection'; and the analysis was generally accepted by the early utilitarians. Whether this is sufficient or not will depend upon the meaning that we read into the term 'reflection', over and above what is contained in 'sensation'; and it is only fair to say that utilitarians like J. S. Mill read considerably more into it than was done by Bentham. But it will depend also upon how we define the individual, who is the subject acquiring knowledge. The old accepted view was to look upon the individual as a self-contained independent unit, bringing with him at birth a mind that is a *tabula rasa*, or that resembles a clean sheet of writing-paper, 'void of all characters', and dependent for all the ideas that he might come to possess on his own experience, learning through personal trial and bungling what was necessary to form character and to make him a success in life. Little or no account was taken either of heredity, or the influence of ancestors upon him; or of the fact that the society into which he is born is organized independently of him and affects him in all-controlling ways through its established customs and institutions, through its prejudices and aspirations, through its limited interests as well as through its ideals; or of the all-important circumstance that he is introduced at birth into a family group, which possesses a formed language, more or less developed, but which, even at its lowest level, imparts to him ideas and knowledge which he himself does not consciously seek, but which he has simply to realize or make his own, yet without which the rapid progress that he makes in intellectual acquisition as time goes on would be impossible. It is altogether ignored that experience could not be to him what it is to the lower animals, or what it would be to a mere individual irrespective of others on whom he is in early life absolutely dependent, and from whom he learns, through speech and otherwise, what he could not himself originate, or what could scarcely even come within the range of his acquisition. In a single sentence, it was forgotten that there is an enormous mass of given social conditions pressing upon him with unceasing force from the very beginning, and ensuring that his special capacities and energies shall act and develop in a particular way and with a speed that is truly amazing. And not only is experience conceived by the utilitarian as the source of knowledge, it is also taken to be the ultimate criterion of truth. It is the final court of appeal, when doubt occurs or dispute arises as to the validity of knowledge— when a demand is made for its justification. Lastly, experience is conceived by the utilitarian as the ultimate source of our moral ideas. Our sense of right and wrong, all that we understand by moral judgement, implicating condemnation or approval of an agent and his action, together with the feelings peculiar to conscience (such as moral indignation and remorse)—in a word, the phenomena of moral consciousness in every form—are all, according to the utilitarian, experiential in their origin and associated with the feelings of pleasure and pain. And not only this, but from experience we get also the criterion or test of moral ideas—the standard by which to estimate their value. The value of a moral principle, it is maintained, lies in the consequences that the application of it entails, i.e. in the amount of general happiness that it can or cannot produce.

It will be evident from this characterization why it is that utilitarianism should have intimate relation with Associationism. By Associationism is meant the attempt to explain philosophically the nature and formation of knowledge and mind out of units of sensation, and an exposition of the principles according to which this formation is effected. It is thoroughgoing in its application, and includes all the processes of the mind—intellectual, volitional, and emotional—and, therefore, claims to be as effective in ethics and in morals as in other provinces of mental science. In the history of philosophy, utilitarianism and associationism have gone together: indeed, the greatest of the utilitarians have also been the leaders of associationism, e.g. Hume, Bentham, the Mills, Bain. The reasons for this are not far to seek:

In the first place, associationism necessarily deals with experience, and utilitarianism is supremely experiential. If it were possible to define happiness apart from experience—still more, if it were possible to secure happiness by means of mere abstract principles supplied to us from without, regardless of experience—associationism would not count. But, if happiness can only be conceived in terms of what man is and what his nature is formed to be, and if it has necessary relations to human wants and aspirations, it becomes needful to discover, in the concrete circumstances of human life, how it is brought about and by what means it can be furthered. This demands a study of how men actually find pleasure and promote their interests, in what ways pleasures are combined and, it may be, transformed, and, accordingly, how association works in riveting and in deepening men's experience. As, moreover, conduct counts most for a happy life, it is necessary to be able to gauge and to forecast consequences of action—to know what a choice of this or a rejection of that is likely to lead to, how the present may tell upon the future, and the like; and this again means association. And association is, further, necessary for the formation of habits (a thing so vital to the ethical man) and for the reformation of the transgressor. Thus, association comes to be of paramount importance, if 'utility' is made the guide of life.

But, in the next place, associationism is necessary, if a scientific explanation is to be offered of conscience (its character, its genesis, its working, its power), and if there is to be rational explanation of the fact that, though pleasure is to be the end at which man aims, he does not always, in the sphere of ethical endeavour, aim at it directly, but makes it his chief business to act in life in accordance with what he conceives to be right and duty. To the utilitarian, as much as to every other serious ethicist, 'duty' stands for the supreme moral idea, with its allies 'virtue' and 'obligation', addressing man with magisterial force; but the utilitarian undertakes to analyse these notions into their constituent factors and to show how they have attained their authority and what a mass of constraining experience lies behind them: and this, once more, means association. Moreover, he strongly insists that this resolution of what is usually taken as ultimate in ethical experience into what is simpler, and building it up again into the formed product, in no way detracts from the value of the product when obtained: on the contrary, it may enhance it, inasmuch as it is now seen to have stood the test of experience and to have been evolved by the race and not merely by the individual.

Then, lastly, associationism comes to the aid of utilitarianism, if the utilitarian ethics is to be scientific. A non-utilitarian ethics may have merits—it may, as the Stoic ethics did, or as Kant's ethics did—appeal to the Puritanic element in man and conduce to an appreciation of the solemnity and seriousness of life; but it can hardly be described as scientific. It appeals only to a part of man's being, and fails to reach the springs of action that flow from the emotional and active sides of his nature. In other words, a merely *formal* ethics is ineffectual for guidance to a warm-blooded social being.

From what has now been said, it is easy to see where the merit of utilitarianism lies. It is intensely human and intensely practical. It is not merely an ethical theory with claims to scientific recognition, but also a theory that enters the realm of politics and aims at finding itself embodied in State legislation. It is directly in touch with the living movements and interests of men, as these are found in society pushing on to a higher level. That, surely, is no mean recommendation.

CHAPTER II

JEREMY BENTHAM: HIS LIFE AND WRITINGS

BENTHAM begins the Utilitarian succession of the nineteenth century, and was the commanding figure of the vigorous English Utilitarian movement. Among the greatest of his followers were James Mill and his son John Stuart, George Grote, and Alexander Bain. Besides being philosophers and thinkers, all these had an aptitude for affairs. Bentham himself was trained to the law, but forsook the Bar for the advocacy of practical legislation and reform. The Mills (father and son) held positions in the India Office; Grote was a banker; and even Bain, who was a university professor, may not inaptly be designated a man of affairs, for, at the opening of his literary career, he was attached to the Board of Health and had a practical training, involving experience in administration, under Edwin Chadwick. On the juridical side, John Austin, in his *Province of Jurisprudence Determined*, developed the utilitarian principles; and Ricardo upheld them in Political Economy. As political thinkers, they were all (with one partial exception) staunch advocates of liberal and progressive measures on a philosophical basis, and belong to the group usually known as 'the Philosophical Radicals'.

We begin with Bentham.

Born on 15 February 1748 in Red Lion Street, Houndsditch, London, Jeremy Bentham lived a full, happy, and laborious life for over eighty-four years, and

died at Queen's Square Place, Westminster, on 6 June
1832. Both his father and his grandfather were lawyers;
and he himself, after graduating B.A. at Oxford in 1763,
at the age of sixteen (completed by the degree of M.A.
in 1766), studied law, and was called to the Bar in 1772.
His inclinations, however, did not lie in that direction;
and so he discarded the idea of practising as a barrister,
and devoted himself to the study of legislation, and
became the strenuous advocate of reform—constitutional,
legal, social, and economic.

As a boy, he was exceptionally precocious, and his
father had high expectations of him as a lawyer, having
visions even of the Woolsack. He was also very sensitive,
especially to two things—blame and fear; and he himself
has told us that his earliest recollection was of 'the *pain*
of sympathy'. That is very significant, presaging the
future champion of the downtrodden and the suffering,
the friend of the lower animals as well as of men. As a
student in his earlier days, while attracted by languages,
he was passionately fond of chemistry and of all experimental science.

Early in life, he expressed himself thus: 'My humble,
but assiduous labours, which I hope will not cease but
with my life, I desire to be engaged in the service of my
country.' So they were; and a noble service did he
render, not likely to be forgotten.

His first literary venture of any magnitude was the
Fragment on Government—an uncompromising attack on
Blackstone's eulogy of the English Constitution. It
appeared in 1776, and brought him immediate fame: in
particular, it attracted the attention of the Whig Lord
Shelburne, then the Secretary of State (afterwards

Marquis of Lansdowne), who, somewhat later (viz. in
1781), after the publication of the *Introduction to the
Principles of Morals and Legislation*, called upon Bentham
and invited him to visit him at Bowood. This was but
the first of many visits paid by Bentham to Bowood,
where he was supremely happy. There he not only
enjoyed the kindness and encouragement of the pleasant
Bowood family, and entered cheerfully into the sociality
of the house (cards, chess, billiards, music, &c.), but
was brought into immediate contact with distinguished
statesmen and eminent men of letters—such as William
Pitt, Camden, Romilly, Dumont, Barré, Dunning.

His writings are voluminous, and, if we include his
correspondence, his published works (apart from his
unpublished MSS.) fill eleven goodly octavo volumes,
closely printed in two-columned pages, in the standard
edition of J. Bowring. The whole constitutes a huge
mass of reasoned matter, well worth study to-day, but
demanding exceptional patience and attention. Some of
the more representative of them, besides the *Fragment
on Government*, are: *A Defence of Usury*, published in 1787;
An Introduction to the Principles of Morals and Legislation,
1789; *Discourse on Civil and Penal Legislation*, 1802; *A
Theory of Punishments and Rewards*, 1811; *A Treatise on
Judicial Evidence*, 1813; *Papers upon Codification and Public
Instruction*, 1817; *The Book of Fallacies*, 1824. In 1824,
also, he founded and financed *The Westminster Review*,
which was destined to play a very important part in
fanning public interest in political questions and in
disseminating Bentham's ideas and principles. In 1827
appeared his *Rationale of Evidence*, edited by J. S. Mill.
His last years were devoted to the production of his

Constitutional Code (part of which was published, before his death, in 1830).

Bentham's literary style changed as the years went on. His early writings are marked by clearness, terseness, and vivacity; but his later works are rendered distinctly prolix and repellent by the over-elaboration of arguments, the excessive love of dissection and detail, and the overloading with technical, uncouth terms—often awkwardly formed and needlessly unattractive. Referring to a published letter of his youth, he himself says: 'Some will say it was better written than anything I write now. I had not then invented any part of my new lingo.' The invented 'lingo' makes all the difference! When, however, we turn to his correspondence, we find that, though his letters to intellectual friends (of various nationalities) dealing with political, constitutional, educational or legislative subjects can be heavy enough, those to others (especially to members of the Bowood household) can be bright and facetious. He can be quite a pleasant correspondent, witty and playful, when he cares. For instance, writing to Lord Holland and discriminating between prose and poetry, he puts it thus: 'But, sir,—oh, yes, my Lord—I know the difference. *Prose* is where all the lines but the last go on to the margin—poetry is where some of them fall short of it.' It ought also to be noted that, although he erred in the exuberance and uncouthness of his terminology, his newly-coined words and phrases are often very felicitous, and that he has enriched the English language with such terms as 'international', 'utilitarian', 'codify' and 'codification', 'maximize' and 'minimize', &c. Further, he had an exceptional power of graphic delineation and of going straight to the point. As an example, we may give his characterization of Samuel Johnson (whether it is just or not must be left to the reader to say) : 'Johnson', he says, 'is the pompous vamper of commonplace morality—of phrases often trite without being true.'

His great influence abroad came comparatively early. It was effected by Etienne Dumont (with whom he first came in contact at Bowood) who reproduced in French and expounded his writings on Legislation, under the title of *Traités de Législation civile et pénale*, published in 1802. France, Russia, Portugal, Spain, and parts of South America fell under its spell. Indeed, so great an influence had Bentham on the French (whose destiny he tried in part to guide at the time of the Revolution) that, on 26 August 1792 the National Assembly conferred on him the title of 'French Citizen'; thereby testifying their grateful appreciation of his efforts in the cause of liberty and the emancipation of nations.

His fame at home moved more slowly; but by degrees it came, till by and by he was acknowledged the moving spirit of a brilliant set of radical politicians, and the high-priest to whom practical reformers in many quarters of the world looked for guidance and suggestion. He was fortunate in having James Mill as his ardent disciple, whose whole-hearted and vigorous advocacy of Benthamism was a potent force in the propagation of it. He had a steadfast friend also in Sir Samuel Romilly, the distinguished lawyer, who served him well; and another auxiliary (though of the younger generation) in Ricardo, the political economist, of whom Bentham used to say: 'I was the spiritual father of Mill, and Mill was the spiritual father of Ricardo: so that Ricardo

was my spiritual grandson.' Among great parliamen-
tarians with whom he got into immediate influential
contact, we find Lord Brougham—whose attitude on
current questions, however, Bentham often criticized,
but with whom, nevertheless, he was on such intimate
terms that he could address him, in the opening of a
letter, as 'My dearest Best Boy', or as 'Dear sweet Little
Poppet', and Brougham made reply to 'Dear Grand-
papa'! In the stalwart Joseph Hume ('that truly honest
and meritorious citizen', as he called him, 'the only true
representative the people of this country ever had, and
one more than, under such a form of government, they
have any right to expect to have') he found an enthusiastic supporter; and no less devoted was Sir Francis
Burdett, on whom he could usually rely and whose
resolutions in Parliament regarding the extension of the
franchise he drafted. Not the least remarkable of all his
adherents in Parliament was Daniel O'Connell. At one
time of his life, Bentham aspired to a seat in Parliament,
and was greatly disappointed at not having his wish
gratified. His hopes arose from a misunderstanding of
a conversation with Lord Lansdowne, whom Bentham
interpreted as promising to find a seat for him.

Bentham had his eccentricities and his peculiarities;
and his hermit life in later years ('the calm of an almost
inaccessible solitude', as Sir Francis Burdett called it)
accentuated his limitations. But his nature was essentially a sympathetic and lovable one. His foibles are
hit off, and his real worth affectionately acknowledged,
in the following sentences from Sir Samuel Romilly's
account of Bentham as host at Ford Abbey: 'We found
him passing his time, as he has always been passing it

since I have known him, which is now more than thirty years, closely applying himself, for six or eight hours a day, in writing upon laws and legislation, and in composing his Civil and Criminal Codes: and spending the remaining hours of every day in reading, or taking exercise by way of fitting himself for his labours, or, to use his own strangely invented phraseology, "taking his antejentacular and post-prandial walks", to prepare himself for the task of codification. There is something burlesque enough in this language; but it is impossible to know Bentham, and to have witnessed his benevolence, his disinterestedness, and the zeal with which he has devoted his whole life to the service of his fellow-creatures, without admiring and revering him.' It stands to Bentham's credit also that he was extremely fond of the lower animals, inveighed against cruelty to them, and advocated legislation on their behalf. He had a cat at Hendon (his 'absconding place') which used to follow him about in the street; he encouraged mice to play about in his study; and he tells us of an interesting friendship thus: 'I became once very intimate with a colony of mice. They used to run up my legs, and eat crumbs from my lap. I love everything that has four legs.' He revelled also in Nature—in flowers and trees and fields; and his garden was a supreme pleasure to him. He not only loved flowers, but set himself to know them botanically, and nothing delighted him more than to get unknown seeds sent him from foreign lands, which he proceeded to cultivate for himself, and to share with his friends.

Bentham was gentle and courteous in his bearing, and considerate of others; yet he could act the Bohemian occasionally. 'Once when Madame de Staël called on him, expressing an earnest desire for an audience, he sent to tell her that he certainly had nothing to say to her, and he could not see the necessity of an interview, for anything she had to say to him. On an occasion when Mr. Edgeworth, in his somewhat pompous manner, called and delivered the following message to the servant, in order to be communicated to Bentham: "Tell Mr. Bentham, that Mr. Richard Lovell Edgeworth desires to see him",—he answered: "Tell Mr. Richard Lovell Edgeworth, that Mr. Bentham does not desire to see *him*." '

He enjoyed life and was naturally cheerful and optimistic, and he retained his youthful spirit to the end. At the age of eighty, he could write: 'I am living surrounded with young men, and merrier than most of them'; and, when at that age also, he invited Daniel O'Connell to visit him, he promised him 'ambulatory conference, for health's sake, in the garden with me', adding immediately, 'Let not the word appal you, for how-much soever your inferior in wit, you will not find me so in gaiety.'

In character, Bentham was one of the most upright and independent of men, scorning to do a selfish or mean thing, and never fearing the face of man. When, instigated by George III, a Declaration was presented to the Court of Denmark, urging a rupture with Russia, Bentham at once gave fierce opposition (this was in 1789), and, under the pseudonym of 'Anti-Machiavel', scathingly analysed the Declaration in several closely-reasoned letters, published in *The Public Advertiser*, holding up the warlike policy to public indignation. An answer to his letters, under the designation 'Partizan', was unsparingly criticized by him. He was led to believe that the answer was the production of George III; and so he said, 'I fell upon the best of kings with redoubled vehemence.' He was firmly convinced that this on-slaught cost him the loss of one of his most cherished projects—his Panopticon scheme—a scheme dealing in a practical way (as we shall see by and by) with the treatment of criminals, which won the sympathy of both Houses of Parliament, but was wrecked (so Bentham believed) by the King. In like manner, when the Tsar of Russia, appreciative of Bentham's desire to supply a Codification for Russia, sent him a gift of a diamond ring, Bentham returned the ring, without even breaking the seal of the packet, so that he might save himself from even seeming to fall under a pecuniary obligation. And, once again, he braved the Duke of Wellington and sternly reprimanded him for his duel with Lord Win-chelsea, in a letter beginning, 'Ill-advised Man!' and ending with, 'Now then, if to personal and physical, you add moral courage, I will tell you what to do. Go to the House of Lords. Stand up there in your place, confess your error, declare your repentance; say you have violated your duty to your sovereign and your country; and promise, that on no future occasion whatsoever, under no provocation whatsoever, in either character—that of *giver*, or that of *accepter* of a challenge, will you repeat the offence.'

Such a man occupied a unique position, and could not fail to be a power in the land. Like many independent men, he was vain and sensitive and could occasionally make himself needlessly disagreeable; but no man repented sooner, or could make the *amende honorable* with greater grace and sincerity.

A trait in Bentham's character was his sympathy with the persecuted and the distressed in the political ferment of the time and his generosity in helping them. Many a person and many a cause benefited by his money in this way. He had the heart to do it, and he had also the means; for he was early in possession of a substantial competence, which was augmented at his father's death, and he remained to the end of his life unmarried.

That consideration for others which characterized him in his earlier days continued to the last. When he felt his end drawing near, he forbade his servants being present, lest they should be pained and subjected to unnecessary suffering. And, choosing his trusted friend and biographer, John Bowring, as his sole attendant at the last moments, he expired with his head resting on Bowring's bosom.

'After he had ceased to speak, he smiled, and grasped my hand', says Bowring. 'He looked at me affectionately, and closed his eyes. There was no struggle—no suffering,—life faded into death—as the twilight blends the day with darkness.'

His body, in accordance with his own instructions, was dissected, in the interest of anatomical science; and the skeleton has been preserved in University College, London, seated in his chair, with the face covered by a wax mask, and wearing Bentham's wonted dress.

CHAPTER III

BENTHAM AS MORAL PHILOSOPHER

In the eighteenth century, English Moral Philosophy showed a great diversity of opinion, more especially on the two questions of the Ethical Standard and the nature of the Moral Faculty. There were the 'Moral Sense' philosophers, like Hutcheson and Shaftesbury, who assimilated conscience to feeling, and maintained that benevolence is the supreme moral principle in man; there were the 'Intuitive' moral philosophers, like Bishop Butler, who erected Conscience into an independent faculty, intellectual in its character, yet operating spontaneously and with a unique authority; or, like Thomas Reid, in Scotland, who appealed to Common Sense (conceived by Beattie as an infallible inner light), whose deliverances were regarded as final, being supported by universal consent, or the acquiescence of men in general; there were followers of Richard Price, who set forth moral perceptions as unimpeachable perceptions of Reason or the Understanding; or of Wollaston, who resolved all into Veracity or the intellectual perception of Truth; there were disciples of Bernard de Mandeville, who insisted that the sole support of virtue is Self-interest. Again, there were Adam Smith, who laid the chief stress on Sympathy as the ground of moral approbation and disapprobation; and upholders of Utility like Hume, and Priestley, and Paley—the last of whom presented ethics in a religious

setting, and defined Virtue as 'the doing good to mankind, in obedience to the will of God, and for the sake
of everlasting happiness'; and there was David Hartley,
whose thoroughgoing Associationism was held to be
sufficient for the explanation of disinterestedness and
conscience, as for other things.

In cognizance of these opinions, although not perhaps
fully versed in each of them, Bentham took up his work;
under the influence of some of them, but vigorously
opposing others.

I. The keynote to his philosophy is found in the
opening sentence of his *Introduction to the Principles of
Morals and Legislation*: 'Nature has placed man under
the governance of two sovereign masters, *pain* and
pleasure. It is for them alone to point out what we ought
to do, as well as to determine what we shall do. On the
one hand the standard of right and wrong, on the other
the chain of causes and effects, are fastened to their
throne. They govern us in all we say, in all we think:
every effort we make to throw off our subjection, will
serve but to demonstrate and confirm it. . . . *The principle
of utility* recognizes this subjection, and assumes it for
the foundation of that system, the object of which is to
rear the fabric of felicity by the hands of reason and of
law. . . . The principle of utility is the foundation of
the present work.'

'By the principle of utility', he continues, 'is meant
that principle which approves or disapproves of every
action whatsoever, according to the tendency which it
appears to have to augment or diminish the happiness of
the party whose interest is in question: or, what is the

same thing in other words, to promote or to oppose that happiness. I say of every action whatsoever; and therefore not only of every action of a private individual, but of every measure of government.' And 'utility' itself Bentham defines as 'that property in any object, whereby it tends to produce benefit, advantage, pleasure, good, or happiness (all this in the present case comes to the same thing), or (what comes again to the same thing) to prevent the happening of mischief, pain, evil, or unhappiness to the party whose interest is considered: if that party be the community in general, then the happiness of the community: if a particular individual, then the happiness of that individual'.

Now, concerning this, it is important to observe that Bentham's doctrine applies, and is intended to apply, not only to morals, but also to legislation; and, as a matter of fact, his great aim was to apply his principles to constitutional, legislative, and law reforms. In other words, he had a living and practical interest in view, and was not merely concerned with barren speculative theory. Hence, he substituted for 'the principle of utility' the more significant phrase 'the greatest happiness principle', or (as he first expressed it) 'the greatest happiness of the greatest number' principle. He is thinking in chief of the good or welfare of the community, and not simply of the individual; but, nevertheless, of the community as composed of individuals, and, therefore, of the individual as one whose happiness is accomplished through co-operation with his fellows. He is testing action and legislation by their effects all round; and through their effects, or the consequences that they entail, must they stand or fall. On the other

hand, that the principle of utility is all-potent is seen
from such facts as these: that men everywhere act upon
it; that even those who criticize it do so on the assumption that it is supreme; and that the two great opposed
principles (a) asceticism, or love of pain, and (b) sympathy and antipathy, or personal like and dislike, are
only the principle of utility wrongly applied. It is to the
principle of antipathy or dislike, oftenest manifested in
mere prejudice, that Bentham ascribes the current
philosophical theories of right and wrong of the intuitional type—theories that he passes successively in
review and rejects.

If, then, pain and pleasure are supreme, it is necessary,
on theoretical and on practical grounds alike, to ascer-
tain the sources of them. This introduces us to Bentham's
enumeration of the constituents of human happiness;
for the meaning of happiness, according to him, is
pleasure and the absence of pain, or the surplus of
pleasure over pain. When alluding to the fact that he
first got sight of the Greatest Happiness principle from
Priestley (he sometimes thinks that he may have got it
from Beccaria—he might have gone to Hutcheson), he
maintains that Priestley, though acknowledging the
principle, failed utterly to realize the true scope and
significance of it, inasmuch as he did not see that the
essence of happiness is pleasure and the absence of pain.
The sources recognized are four in number: the physical,
the political, the moral, and the religious. Each of these
is a 'sanction', inasmuch as the pleasures and pains
belonging to it give a binding force to any law or rule
of conduct. When pain or pleasure comes to us in the
ordinary course of nature, without any intervention or

purposeful modification of will, it is said to issue from the *physical* sanction, e.g. temperance conserves health and produces pleasure; disease is naturally brought on by intemperance, and pain is the result. When it comes through properly constituted authority in the community, and is administered by a particular person or persons duly accredited (say, a judge), it issues from the *political* sanction, or what we usually know as the law of the land. The *moral* sanction designates the pressure of public opinion upon us, and should more properly be called the *popular* sanction. The *religious* sanction has reference to our belief in God and His relation to us in the present life and for the future.

The problem for the moralist and the legislator, then, is apparent. It is how best to make these sanctions operative for human happiness, individual and general; the moralist and the legislator being both actuated by the same motive, though each having his own method.

But pleasures and pains differ, not only as to their source; they differ also as to their worth or value. And so we must next determine the mode of measuring 'the value of a lot of pleasure or pain'. This is clearly important from the standpoint of the legislator, whose chief concern is the apportioning of lots of happiness, or, at any rate, legislating in such a way that happiness may be distributed in the community on the principle that 'everybody is to count for one, and no one for more than one'. But it is indispensable also for the moralist. How, then, are we to estimate value for the individual, and how are we to estimate it for numbers of individuals, or for society in general?

So far as the individual is concerned, the value of a

pleasure or a pain, considered by itself, depends on four things—its intensity, its duration, its certainty or uncertainty, and its propinquity or remoteness. If, in addition to estimating the value of a pleasure or a pain taken by itself, we wish to estimate the tendency of the *act* that produced it, two other considerations have to be taken into account, viz. its fecundity (i.e. the likelihood of its being followed by sensations of the same kind, pleasure by pleasures and pain by pains) and its purity (i.e. the likelihood of its not being followed by sensations of the opposite kind, pleasure by pains or pain by pleasures).

So far as a collection of individuals is concerned, not only have all these six circumstances to be taken into account (intensity, duration, certainty or uncertainty, propinquity or remoteness, fecundity and purity), but also a seventh, viz. the extent of the pain or pleasure, that is, the number of persons affected by it.

So then, it is a matter of a hedonistic calculus—of summing up pleasures and pains in any particular case, and balancing the pleasures against the pains, and estimating the value accordingly. This is the theoretically perfect process; but, in actual practice, in a civilized society like that enjoyed in Great Britain, it is not necessary to go through the process strictly, previously to every moral judgement formed or to every legislative or judicial operation. Things are shortened for us by the fact that we live in an organized community, with customs, laws, rules, and institutions provided for our guidance, based on a large and varied experience.

But more even than this is necessary, if the utilitarian

principle is to be sufficient to explain both moral and political action—action with a view to legislation: there is needed a distinct enumeration of the kinds of pleasures and pains. This, accordingly, Bentham offers. After distinguishing between simple and complex pleasures and pains, he devotes considerable space to the elucidation of those of them that are simple, setting down the pleasures as fourteen and the pains as twelve. The simple pleasures are those of sense, of wealth, of skill, of amity, of a good name, of power, of piety, of benevolence, of malevolence, of memory, of imagination, of expectation, of association, and of relief. The simple pains are: pains of privation, of the senses, of awkwardness, of enmity, of an ill-name, of piety, of benevolence, of malevolence, of memory, of imagination, of expectation, and of association. It is obvious that this enumeration of pleasures and of pains is not made on any scientific or logical plan: it is not exhaustive, nor are the members of it (in either case) mutually exclusive. It is simply a rough collection, adequate, perhaps, to practical purposes.

In this connexion, we may advert also to the importance, both for the moralist and the legislator, of paying regard to the circumstances that influence sensibility. These circumstances Bentham gives as thirty-one. We need not follow him. But they are such as health, bodily imperfection, quantity and quality of knowledge, strength of intellectual powers, bent or inclination, moral and religious sensibility and biases, pecuniary circumstances, rank, education, Government, &c.

The way is now clear for a consideration of Human Action in general; dealing with the distinctively ethical

notions of right and wrong, good and evil, merit and
demerit as attaching to actions, and with the nature of
punishment and the apportioning of it by the moralist
and the legislator respectively.

In estimating the morality of an action, as also in
dealing with a particular act of legislation, the intention
of the doer has to be taken into account; and we have,
further, to take account of his consciousness of consequences. But not intention only, illuminated by consequences, has to be considered; it is necessary, in
addition, to consider motive. The two things, intention
and motive, are by no means the same. For example, in
doing a particular action, a man's intention may be (say)
to benefit a neighbour, but the motive that urges him
to it is the particular regard that he entertains toward
that neighbour, or it may be his regard for some friend
who has requested his good offices in the neighbour's
behalf. The motive is what prompts him to act; the
action itself is covered by his intention. How, then, does
motive stand related to intention?

In handling this question, Bentham deals with
intention just in the way that we might expect a philosophical lawyer to do, drawing the distinction between
intending an act and intending its consequences. It is
indisputable that, in apportioning blame or responsibility, much will depend on what exactly the agent meant
to do. If, in playing a practical joke on a person, a
man, without intending it, also injures the person, the
gravity of the situation is mitigated by the fact that the
injury was unintended. On the other hand, if a man
intended to injure another, but his act miscarried and
no injury was done, he cannot claim exemption from the

guilt of the intention, even though the person aimed at were unharmed.

All this is plain enough from the side of the moralist, but there are obvious difficulties in carrying it out by the legislator and the judge, for whom the overt action or the consequences must count for most. In the eye of the judge, in a Court of Justice, the accused has either done or not done the action laid to his charge, and according to the evidence he is acquitted or condemned.

With regard to motives: they are what prompt, induce or determine the will; and, in the ultimate analysis, what so prompts and determines are pleasure and pain. So that 'a motive is substantially nothing more than pleasure, or pain, operating in a certain manner'. Moreover, it is a point that Bentham insists on with great rigour that motives are not in themselves either constantly good or constantly bad, but that a motive is only good or bad 'with reference to its effects in each individual instance; and principally from the intention it gives birth to: from which arise . . . the most material part of its effects. A motive is good, when the intention it gives birth to is a good one; bad, when the intention is a bad one: and an intention is good or bad, according to the material consequences that are the objects of it.'

Note may here be made of the fact that, with his usual thoroughness, Bentham drew up an elaborate table of the Springs of Action, the nature of which may be sufficiently gathered from the long title that he gives to it, in which also we see his pedantic love of technical terminology: 'A Table of the Springs of Action: showing the several species of pleasures and pains of which man's nature is susceptible, together with the several species of Interests, Desires, and Motives respectively corresponding to them: and the several sets of appellations, Neutral, Eulogistic, and Dyslogistic, by which each species of motives is wont to be designated: to which are added Explanatory notes and Observations, indicative of the applications of which the matter of this Table is susceptible, in the character of a basis or foundation, of and for the art and science of Morals, otherwise termed Ethics, whether Private or Public *alias* Politics (including Legislation)—Theoretical or Practical *alias* Deontology —Exegetical *alias* Expository (which coincides mostly with Theoretical) or Censorial, which coincides mostly with Deontology: also of and for Psychology, in so far as concerns Ethics, and History (including Biography) in so far as considered in an Ethical point of view.'

What, next, let us ask, is the order of pre-eminence among motives? It is determined by the greater or less likelihood of its dictates, taken in a general view, being coincident with those of the principle of utility. That being so, Goodwill manifestly takes the first place. This is the principle of Benevolence, of which British moralists of the eighteenth century and the early part of the nineteenth century made so much. The difficulties connected with the application of benevolence arising from the competition of consequences, according as the benevolence extends to a larger or a smaller group of human beings, and according as the extent of interests conflicts with their importance, must be directly faced. This means that benevolence, in order to be effective, must be both extensive and enlightened—a restricted benevolence may err. Matters are eased to us in the actual working by the fact that the dictates of private benevolence rarely conflict with those of public benevolence.

Next to Goodwill in order of pre-eminence comes Love of Reputation. Here, too, the dictates, for the most part, are coincident with those of public utility. When the coincidence is disturbed, it is mainly owing to the fact that people allow themselves in their likes and dislikes, in their approbations and disapprobations, to be guided, not by utility, but either by asceticism or by sympathy or antipathy.

And so with the other two principles that operate as motives, viz. Desire of Amity or personal affection, and the Dictates of Religion. Each of the two is determined in the order of pre-eminence by the tendency of its dictates to coincide with, or to frustrate, those of the principle of utility; and the order is—first the desire of amity (or personal affection), and next the dictates of religion. Religion comes last in Bentham's estimate because of the different and often conflicting notions of it by the different denominations and the unlikelihood of finding a general agreement.

This exposition and grading of Motives is of the highest importance for Benthamite philosophy. By placing Benevolence at the top, and by appraising the whole system by the test of the greatest happiness of the greatest number, not forgetting that the individual is to count for one, it renders nugatory the objection that has so frequently been brought against utilitarianism that it is essentially a selfish system. Universalistic hedonism is anything but selfish, even although the individual's pleasure may be at the root of it. Self-love and selfishness are by no means the same thing.

We have just seen that good and bad are not predicates strictly applicable to a man's motives. What, then, is there about him to which these predicates may be properly applied? To this the answer is: His disposition. 'Now disposition is a kind of fictitious entity, feigned for the convenience of discourse, in order to express what there is supposed to be *permanent* in a man's frame of mind, where, on such or such an occasion, he has been influenced by such or such a motive, to engage in an act, which, as it appeared to him, was of such or such a tendency.' What, then, determines the goodness or badness of disposition? Just, as in the other cases, its effects—its effects in increasing or diminishing the happiness of the community, including that of the individual himself.

Here, it is necessary to observe that disposition is ultimately associated with intention; and two things are of vast significance, both borne out by our experience —(a) that, in the ordinary course of things, the consequences of actions usually turn out conformable to intentions and (b) that 'a man who entertains intentions of doing mischief at one time is apt to entertain the like intentions at another'.

What, now, of Punishment? So far as politics and jurisprudence are concerned, the answer to this will meet us later on; but, so far as concerns morals, the answer may be given now. In the first place, punishment should not be vindictive: it must not be inflicted merely for the purpose of giving pleasure or satisfaction to the person injured or aggrieved, although this may be a collateral end served by it. As a true utilitarian, Bentham recognizes the pleasure of revenge, and of the malevolent affections generally, as native to human nature, and requires that it shall count for its worth in determining the happiness of the revengeful individual. Of vindictive satisfaction he says very explicitly: 'This pleasure is a gain: it recalls the riddle of Samson; it is the sweet which comes out of the strong; it is the honey gathered from the carcase of the lion. Produced without expense, net result of an operation necessary on other accounts, it is an enjoyment to be cultivated as well as any other; for the pleasure of vengeance, considered abstractly, is, like every other pleasure, only good in itself. It is innocent so long as it is confined within the limits of the laws; it becomes criminal at the moment it breaks them.' In the next place, punishment should aim at the amendment or reformation of the individual transgressor. This removes the apportionment of punishment from the wish or desire of the avenging individual to the realm of reason and the fact of the solidarity of the race, or the natural responsibility of a man for the welfare of his fellow. And, in the last place, punishment should have in view the effect, by way of example, upon the community—in other words, its effect should be prohibitive or deterrent.

This summary of the salient ethical positions of Bentham, following mainly the treatise on Morals and Legislation, may serve to show the thoroughgoing and insistent way in which he carries his ruling principle, the Greatest Happiness principle, through the various spheres of moral conduct. He regards the essence of happiness to be pleasure and the absence of pain, and claims that, although he got the suggestion of the principle from Priestley, he made an entirely new and original use of it, by thus seizing the essential point of happiness and carrying it out in all the details of its workings. It became necessary also to connect pleasure with its springs, so as to give its ethical and moral bearings. He has, further, a scheme of pleasure values; estimated, however, according to the quantity of pleasure, without consideration of its quality.

A brief survey like this of a huge mass of material is necessarily somewhat dry and lifeless; but there is life enough in Bentham's working out of his thesis, and he has the great merit of engaging the reader's attention by the frequent use of concrete examples and happy illustrations, thereby making him feel that it is no mere academic discussion that he is listening to, but that a real effort is being made to meet the living man and his difficulties and to help him to understand the immensely important subject discussed, with a view to his own life and conduct.

II. No sooner were Bentham's views given forth to the world than the critic and the objector appeared, and they have been at work ever since. One or two objections aimed at the doctrine of pleasure may be considered.

It has been urged that if, as Bentham maintains, the ultimate motives of human action be pleasure and pain, then these are to be measured only by their quantity and by the limits which the physical organism places to the realization of pleasure or the experience of pain. This means that we are reduced to shrewd calculation of the results that indulgence or restraint is likely to produce, and Prudence becomes the ruling virtue. In other words, Bentham's philosophy rises no higher than that of self-regard, which, the objector says, is but another name for selfishness, and is incompatible with lofty ethical aspiration and achievement.

But the objection in this form is not decisive; for prudence is not by any means identical with selfishness, nor is it to be treated as a despicable virtue. On the contrary, given man's dependence on the body as a sentient organism and the limitations to enjoyment which that dependence implies, together with the native tendency in human nature to go beyond the limitations, and prudence becomes an important virtue, of high significance to the legislator. Where Bentham fails is, not in setting value on prudence, but in not sufficiently emphasizing the fact that the individual's prudence is *socially* conditioned, and so is inseparable from the welfare of others. He does not appreciate pure disinterestedness, but ultimately resolves it into pursuit of individual pleasure. We do a disinterested action, he holds, because it gives us pleasure, or because the doing of it frees us from a pain which would be greater than the pleasure that the doing of it brings. This is certainly to lower the character of disinterestedness, and to ignore patent facts in our ethical experience. Some later utilitarians, such as Bain, set themselves to rectify this defect.

Again, an objection to Bentham has been raised on the ground that consideration of pains and pleasures does not give us morality at all, but only sentient experience. In answer, it may be said that, although it is quite true that pleasure *as* pleasure is neither moral nor immoral, yet, inasmuch as the tendency of pleasure is to transgress bounds or to go to excess, it needs to be placed under the control and illumination of reason; and, whenever reason's control comes in and man's appetites and passions have to be restrained and the present gratification has to be forgone because of future consequences, morality emerges. In other words, selection among pleasures and moderation of pleasures in general have to be exercised in the view of consequences, to the individual and to others; and that means morality. Morality is essentially rational control; and it would never come into view at all if pleasures did not compete and over-indulgence lead to serious results. This is accentuated by the fact that the individual is a 'person' among other persons, and that his pleasures have to be limited by theirs.

Plausibility is given to the objection, perhaps, by the fact that the Benthamite is assumed to be a man constantly concerned with a cold, selfish, brooding-over of results—that he can never move or act until he has first calculated, deliberately and consciously, how the particular movement or action is to turn out. But this constant conscious calculating process is not demanded. Bentham insists that, in a civilized community, conduct is moulded for us in large measure by convention and society, and that social rules are generated and laws enacted embodying results of action as they have been crystallized by centuries of experience, so that individual deliberate calculation is rather the exception than the rule. It is only seldom that we need to sit down and laboriously work out the consequences for ourselves. In ordinary cases, we act with all the spontaneity and non-conscious readiness that habit, custom, and acquiescence in the collective wisdom produce, and that the moral man is said to exhibit as the very essence of his moral character. Only occasionally do we need to reflect upon and justify moral action; and then, when this happens, the final appeal is made to consequences, interpreted as happiness—adding to or conserving our pleasures, or else detracting from them or substituting pains for pleasures. An end may be effectively aimed at without the individual having it, moment by moment, in conscious view; and definite consciousness of it is least necessary when men are living in a highly-developed social state, where moral conduct is consolidated and the members are the heirs of the ages.

A further objection has been made on the ground that Bentham insists on testing conduct by the number and quantity of the pleasures that it produces; but that is an impracticable test, inasmuch as 'a sum of pleasures' is an impossible conception.

This objection would have force if ethics were an abstract science, strictly mathematical and demonstrative in its character—a science where absolute exactitude of measurement were possible at all points, and where psychology and experience were ignored. But it is not valid when we are dealing with a practical science like ethics (or politics), where mathematical precision is impossible, but where, nevertheless, experience guides and the motives by which men are prompted may be approximately discovered and their results approximately foreseen. From the necessities of the case, we cannot know the *whole* of a man's character or motives—

not even does the individual know the whole of his own character or motives; but we have, nevertheless, to act and to judge on the knowledge that we possess. And the nature of the case also requires us, in forecasting results, to work by consideration of *tendencies*—which must often be merely guessed at, or, at the best, appraised by our knowledge of averages. Moreover, there is no incompatibility between aiming (not necessarily with full consciousness) at a sum of pleasures, or pleasure on the whole, and (say) acting or thinking for the sake of action or thinking, without an immediate reference to self. This has been greatly misunderstood. So long as intellectual contemplation and disinterested conduct are inseparably associated with or accompanied by pleasure (which in our experience they are), they must enter into the calculation of 'a sum of pleasures', for a conscious state is not simply pleasant, but pleasant as modified by the other contents of the state; and although pleasure, on any given occasion, may not be the end consciously aimed at in the action, or the thing that we are immediately seeking in intellectual contemplation, it is the practical test by which we gauge the desirability and the significance of the action, and which affords us the reason why we go on thinking or why we devote ourselves continuously to contemplation. A 'sum of pleasures'—such as we explicitly formulate— may not adequately represent the whole situation, if we demand mathematical precision and exhaustive analysis; but it is the best practical standard by which we can weigh and measure it. And, as a matter of fact, it is a standard that men constantly employ. Whatever other things they may aim at or desire (so experience teaches), they desire to have as much pleasure as possible, pleasure as fully organized as possible, and as long a time of enjoyment of pleasure as possible. They have an idea of the fullness of pleasure, which captivates their desires, and to this they add continuance or repetition; which two things constitute a sum (fullness *plus* continuance or repetition) that is operative in determining their conduct.

CHAPTER IV

BENTHAM AS SOCIAL AND POLITICAL THINKER

I. GENERAL POSITION.—When Bentham began to write on political questions, it was the moment of insistence on 'the natural rights' of man. The Revolutionists in France had made this the basis of their claims; and the Americans had done the same, in their Declaration of Independence. The doctrine had found staunch up-holders in England in Tom Paine and Godwin. It was strenuously opposed by Bentham. He called natural rights 'simple nonsense: natural and imprescriptible rights rhetorical nonsense—nonsense upon stilts'. His reasoning has been patly put by Sir Leslie Stephen in this way: 'The "rights of man" doctrine confounds a primary logical canon with a statement of fact. The maxim that all men were, or ought to be, equal, asserts correctly that there must not be arbitrary differences. Every inequality should have its justification in a reason-able system. But when this undeniable logical canon is taken to prove that men actually are equal, there is an obvious begging of the question. In point of fact, the theorists immediately proceeded to disfranchise half the race on account of sex, and a third of the remainder on account of infancy.' What rights a man has are not 'natural', but, according to Bentham, such as are given or allowed him by law; and, as the worth or goodness of the law itself is its utility, the degree in which it conduces to the greatest happiness of the greatest number, the theory of natural rights is replaced by the theory of utility.

In like manner, Bentham rejected the theory of Blackstone, who, following earlier writers, based political obligation on a primitive social contract. There is no evidence that such a contract ever existed; but, even if we suppose that it existed, the question is not settled. For we immediately go on to ask, Why was such a contract necessary—what is the end that it serves? and, Why should a man keep a contract? To this there is only one satisfactory answer, in the view of Bentham— Utility, or the general good.

II. THEORY OF GOVERNMENT.—Such being Bentham's view of political obligation, we are now prepared for his speculations on Government, and his drastic practical proposals for reform.

Starting with the fact of Representative Government or government by the majority of representatives duly elected by the people, he set himself to consider how best such government might be carried on, and what reforms would be necessary in the British Constitution for that end. For he was far from considering 'the matchless constitution' as perfect. Three things in particular he counselled with a view to amendment. First, *Universal manhood suffrage*—subject, however, to the condition that the adult exercising the franchise should be able to read. This qualification was in the interest of education, which Bentham (and all the utilitarians) greatly valued. On the other hand, he eschewed the question of women suffrage; brushing it aside with the reflection that it would be time to consider it when there was a real demand for it. The demand in his day came from an insignificant number, but the opposition was exceedingly strong; and, in the circumstances, he held that submission to the few would be so obvious an annoyance and injustice to the many that the thought of it might be at once dismissed. Secondly, *Annual Parliaments*. The brief year's duration of a parliament appeared to him to give security against self-interest and lethargy on the part of the members elected. But it would also go far towards securing that the legislator keep himself in constant touch with his constituents, and afford an opportunity to the electors to judge their representative, should he show a tendency to hold and enunciate views opposed to theirs. Thirdly, *Vote by Ballot*. This is required in the interest of electoral purity—a safeguard against intimidation and bribery. In this we have a point thoroughly characteristic of the utilitarians, though, as we shall find, J. S. Mill was opposed to it. Its most strenuous advocate in Parliament, later on, was George Grote.

Things have moved far since Bentham's day, yet in his direction. The Ballot is an accomplished fact, and Manhood Suffrage seems coming within the range of practical politics. Only the proposal for annual parliaments has been dropped, and is not likely to be revived. The need for it is gone. There is no longer the difficulty of sufficiently frequent communication between a member of Parliament and his constituents living a long distance apart. Railways and motor cars and steam-boats, the telegraph and the telephone, not to speak of the penny postage, have brought the representative and his electors, however far separated in space, very near together; and there is the Press, with its eye on parliamentary members, and keenly canvassing political opinions at all points—the importance of which Bentham so fully appreciated that he stood forth as a champion of the Freedom of the Press.

The object of all these proposals was to secure the real and effective representation of the people: the democracy must have its full weight. For this purpose, still another thing seemed necessary—the equalizing of electoral districts. So long as inequalities remained—small constituencies here, large constituencies there—bribery and corruption would go on, accentuated in the case of the smaller constituencies because of the comparative paucity of electors and the facility of concentrating upon them, imperilling the benefits of secrecy secured by the ballot.

But even greater reforms than these were contemplated. Bentham intensely disliked the hereditary character of the House of Lords, which he regarded as having no defensible foundation. But he was also strongly opposed to a Second Chamber altogether: he would sweep it clean away, and leave only the one legislative chamber of the people's representatives. In that event, his proposal of annual parliaments (with provision for carrying forward legislation from one parliament to another) came to his aid, offering security for speedy legislation and the efficiency of the members.

But he went a step farther still, and assailed the Monarchy itself. He had no love for kings, and he had unbounded dislike of George III, and spoke of him in very uncomplimentary terms. Indeed, he held him up to public scorn, and especially to the scorn of the French in his *Jeremy Bentham to his Fellow-citizens of France, on Houses of Peers and Senates*. His faith lay in a Republic. In that direction, he thought, might be found both efficiency and economy, and the supremacy of the people. On one thing he was clear, that the interests of monarchs were not identical with the interests of their subjects, and that the enormous expense of a monarchy, with only obstruction to popular legislation in return, was money ill spent. Hence his active sympathy with France and the French Revolution, and with the United States of America and their Independence. Hence also his spirited defence of the people against the current charges of self-interest and the desire to overturn the principles of justice and common sense. The whole doctrine of his *Constitutional Code* has in view a republic; and he himself, writing to Admiral Mordvinoff, in 1824, declared its object to be 'the bettering of this wicked world, by covering it over with Republics'.

This extreme Radicalism may seem surprising to come from one who opposed the doctrine of 'the natural rights' of man. But he felt it to be only the logical outcome of his leading principle of maximum happiness. Given a monarchy, he reasoned, and the King's interest alone is supreme; given a limited monarchy, and the interest of a privileged class, as well as that of the sovereign comes in; it is only when democracy rules that the interests of the governors and the governed become identical, for the greatest happiness of the greatest number is then the supreme end in view.

III. LEGISLATION.—We have already seen, in Chapter III, the intimate relation, in Bentham's view, between legislation and ethics, or, as he called it, in a term of his own coining, Deontology (the science of right and duty). They both deal with human actions and aim at directing them; and they are both concerned with determining and apportioning human happiness. Yet, Ethics has for its object the guidance of the individual, or how he should frame his life and mould his character; and Legislation is concerned with what is proper to be commanded and enforced, in the interest of public welfare. These two things, though allied, are not the same; and the one sphere is narrower than the other. To make laws for the land enjoining things to be done (or forbidden) and enforcing the command by pain or a penalty is one thing (there is threat or coercion involved); it is quite another thing to appeal to a man as a free-will agent, and get him by sweet reasonableness cheerfully to adhere to a principle, or to pursue a certain course of action. It is the difference between 'must' and 'may'—between the unqualified imperative commanding and the realization of the capacity of one's nature to identify duty with love and to accept the right and render service spontaneously and freely. Hence the difficulty that confronts the legislator. As it has been put by Hill Burton, summarizing Bentham, 'that which it may be each man's duty to do it may not be right for each legislator to enforce upon his subjects, because the very act of enforcement may have in it elements of mischief to the community, preponderant over the good accomplished by the enforcement. In other words, it may tend to the greatest happiness of society, that a man should voluntarily follow a certain rule of action; but it may be injurious to the happiness of the community in general to compel him to follow such a rule if his inclination be against it. For instance, in the *Defence of Usury*, the lending and borrowing of money at high interest, for the purpose of improvidently ministering to extravagance, is condemned; but, on the other hand, it is found that the laws for suppressing usurious transactions are so mischievous in their effect that they too are condemned for precisely the same reason—their malign influence on human happiness'.

The legislator has difficulties also in adjusting the various ends that the law has in view. These ends are four in number—security, subsistence, abundance, and equality; and they are to some extent conflicting. The difficulties, if acutely felt in Bentham's day, are no less acute still. Questions such as these arise and press for a solution: Are people to be allowed to starve before the means of those who have plenty are interfered with? What of the land problem and security of property? What about Labour and strikes and the claims of Industry and the shaking of public confidence? Clearly, the difficulties in matters such as these are but instances of the one great difficulty of how to adjust the various antagonistic aims and claims, and make the sacrifices that are necessary, on this side and on that, if even a tolerably satisfactory reform is to be effected. How is equality to be adjusted to abundance? How is abundance possible without security? What can be of any worth, if subsistence be wanting?

Next, as the object of legislation is the good of the people, and as laws are made to be obeyed and not to be broken, it is necessary for legislation to carry the people along with it. No doubt, laws may be enforced (and in some circumstances should be enforced) whether they are popular or not; but it is only when people voluntarily acquiesce in them, and accept them without coercion, that they can be truly effective. It is this general acquiescence that gives to legislation its permanence and efficiency, and makes it conduce to the happiness and welfare of the community. General dissatisfaction means ultimately rebellion. Therefore, in order to secure a ready acquiescence on the part of the community, the reasons for legislation should be given and made plain and obvious.

The practical reforms advocated by Bentham are too numerous to mention. One was the reformation of the Poor Laws, on the guiding principle of utilizing the able-bodied pauper and suppressing the mendicant or 'sturdy beggar'. In this relation, he was the first to sketch a system of education for pauper children, and to suggest the institution of 'Frugality Banks', which has developed into the 'Savings Bank' system of to-day, which is now such a power for good in the land. Again, he turned his attention to Health, and made proposals which were by and by carried into effect by Edwin Chadwick, head of the Board of Health, and which have assumed the magnitude and importance of the Sanitation legislation of the present day. On every side, his ideas overflowed; and they were mostly of the practical kind, which has told in later legislation.

IV. POLITICAL ECONOMY.—Like most other thinkers of the time, Bentham was an ardent follower of Adam Smith; but he did not hesitate to dissent from his master at points where he thought that Smith had erred. For instance, accepting the position that Government ought not to interfere unnecessarily with the law of supply and demand and that it should allow the greatest possible liberty to the individual in his dealing with his fellows, he rejected Smith's adherence to State legislation against Usury, regarding this as a lapse from his own principles, and upheld the doctrine of non-interference. This is the subject of his little treatise on *The Defence of Usury*. The title is rather misleading. The book is no defence of usury in the sense that it supports the usurer or defends his practice, but is simply an exposition of the position that it is unwise of the legislator to interfere with the usurer, inasmuch as interference is certain to do more harm than good.

Needless to say, Bentham was a strong adherent of the doctrine of Free Trade. He works out the subject with great fullness, laying down principles and meeting objections, and illustrating all out of his abounding knowledge; and he lived to see his principles on the point of realization. In his *Rationale of Reward*, he lauds the principle of unlimited freedom of competition, showing the many advantages that accrue from it and the many disadvantages that a limitation of free competition entails. All limitations, he holds, are just so much injury to the national wealth. Only by free competition is it possible to secure the lowest prices and the best work, and also to make sure that the most vigorous and enterprising shall prevail. As trade is the child of capital, he has much to say on the relation of capital to trade; and his thoughts on the subject are worth considering at the present moment.

Bentham had no particular liking for the Colonies. Although the retention of them might in some ways conduce to the welfare of the Colonies themselves and to the good of mankind, he regarded them as being far from a source of wealth to the mother country, and he would have let them go without compunction; taking care, however, that no new ones should replace them. His position was, that possession of the Colonies is not necessary to carrying on trade with them, and that, even when trade is not carried on with them, the capital that such trade would have required might be applied as productively to other undertakings. This was a doctrine that appeared again and again in the Utilitarian School. The moment of Imperialism was not yet; nor could the magnificent support of the mother country by the Colonies, in men and money, in the hour of danger, as at the present moment, have been foreseen.

How Bentham contended against monopolies, bounties, and the like, is matter of past history, and need not be enlarged on here. His great object was to expose at all points 'the fallacy of those artificial efforts which legislation makes to increase the country's wealth'; and, in large measure, he succeeded.

V. EDUCATION.—Like other great reformers, Bentham had unswerving confidence in the power of Education to procure happiness for the individual and efficiency in his work, and also to improve the race. Hence, he advocated a system of National Education, and required in a man ability to read as an indispensable condition of his exercising the franchise, and he provided for the education of the criminals in his Panopticon. But he went farther and drew up two allied schemes—one appropriate to the poorer or lower classes, and the other to the middle or upper ranks of society. The first has special reference to the education of pauper children, who seemed to him to claim the attention of the State in a very special degree; and it was set forth in connexion with his critical treatment of the Poor Law and its administration. He wished to raise these unfortunates out of the grade of outcasts, to which they had been hitherto condemned, and to fit them for being good and profitable subjects of the King. For this purpose he urged the necessity, first of all, of laying in them the foundation of good habits, which could not be done apart from moral teaching. It was, in the first instance, a matter of personal character, which needed social intercourse as well as verbal inculcation of principles, practice no less than instruction, to make it stable and satisfactory. There next was needed training in other ways suitable to the circumstances. As being indigent children destined to make their living by some kind of manual labour, it was necessary to instruct them in a trade or means of livelihood. But, further still, there must be improvement of the mind—intellectual instruction, which should arouse and develop the mental faculties, and produce for the individual a permanent source of pleasure and of power. Thus were pauper children to be raised in status and equipped for playing an independent and a worthy part in life. In all this, Bentham was before his time. For we must be careful not to read back into his day the widespread interest in the education of the masses that characterizes the present age. On the contrary, there was little enthusiasm for general education then. Legislators and the ruling classes were afraid to educate the people, lest education should prove a danger to society; and they grudged the expense. Distrust of the people and selfish economy combined to maintain the existing order of things.

But Bentham legislated also for the education of the wealthier (but not the professional) classes, or, as he called them, 'the middling and higher ranks of life'. His scheme was an adaptation and extension of a new system of instruction that had just been introduced in the end of the eighteenth century. Among educationists of the time, two names stand out conspicuous: those of Dr. Andrew Bell and Joseph Lancaster—the one a Scotsman, the other an Englishman. The innovation consisted in the attempt to awaken the spirit of unity and the feeling of corporate action in a school by introducing the Monitorial system, i.e. by utilizing the older or more advanced pupils in instructing the younger and less advanced. In Scotland, this was applied with great success in the teaching of Latin and of Greek by Dr. Pillans in the High School of Edinburgh. But it was Bentham's distinction to give the system a much wider range. He embodied his views in his *Chrestomathia*. The name is significant. It is a coinage of his (though he afterwards found that he had been anticipated), compounded of two Greek words, signifying 'useful learning', or 'the study of useful things'; the emphasis being laid on the epithet *useful*.

The principles on which his Chrestomathic scheme proceeded were these: First of all, it aimed simply at intellectual instruction, deliberately omitting morals and religion. This differentiated it from Lancaster's system and from Bell's; the former zealously inculcating Scripture knowledge, but in an unsectarian fashion, and the latter strictly adhering to the religious doctrinal teaching of the Church of England. His position needed explanation; and so he opened his *Chrestomathia* with a detailed exposition of the value and utility of learning or instruction of the intellectual stamp. Next, his system of teaching started from the position (by no means self-evident at the time he wrote): Begin with what is useful—what is most likely to be of service to the pupil in his after career in life. This was, to a certain extent, a revolt against the dominance of classics in the school and college education then in vogue. Bentham had personally no ill-will to classical learning—he was himself an excellent Greek and Latin scholar, and made ample use of his Greek knowledge in coining his copious technical vocabulary; but he felt that for, say, a member of Parliament, a preliminary school training on other lines than that of the dead languages would be far more effective. He had great faith in bringing the young mind into immediate contact with Nature and natural science, and of creating in it an interest in the structure, processes, and phenomena of the wonderful world in which it was called upon to energize. For the same reason, he included in his scheme 'useful skill', as well as intellectual knowledge; and he duly recognized the educational value of modern languages. It is interesting also to note, in view of recent developments, that, in his scheme, he early begins the pupil's instruction with nature-knowledge subjects, such as botany and zoology.

Had he lived to-day, he would have advocated also local history and local archaeology at least. Thirdly, a peculiarity of the system lay in due attention being paid to the grading of subjects, with reasons adduced for the order of succession, starting with the principle: Teach first the things that are easiest to learn, i.e. pay regard to the learner's capacity, and do not force him contrary to his aptitude and his natural inclination. The whole end of the scheme was to widen the pupil's knowledge, to enlist his interest, and to broaden his sympathies, so as to enable him to get the most out of life, and to fit him for being a worthy citizen.

Equal attention did Bentham pay to the problem of School Management. In this connexion he works out with much elaboration the monitorial system, and excludes from school discipline, contrary to the practice of his day, corporal punishment.

All this is extremely important, and shows that Bentham's views were precisely in the line of the future development of education in Great Britain. They were also a force in setting going that marvellous educational progress that has taken place since his day—from the monitorial system to the pupil teacher system; from that to Normal Schools, devoted to the actual training of teachers—theoretical and practical; to be in turn supplanted, at the present time, by the vast State-aided organization of Training Centres, with Provincial Committees, under the authority of a Central Board or Education Department. All this has only to be translated into the usage of other English-speaking nations—the United States and Canada—to see how much education has owed to Bentham.

CHAPTER V

BENTHAM AS JURIST AND LAW REFORMER

LAW REFORM; PUNISHMENT AND PRISONS

I. LAW REFORM.—Of the many needs of Bentham's time, there was none more clamant than that of the reform of the law; and he bent his mind towards it with characteristic energy and determination. His sympathy for the people also, and his eagerness to see justice administered and happiness secured to the deserving and the oppressed, explain his zeal and, in part also, his success. His writings on the law, or on questions relative thereto, fill a very large space. They are occupied in great measure with criticizing existing laws and the existing machinery for the execution of them, or proposals for new laws of which he did not approve. Common law, statute law, law in all the forms that are known in England, were surveyed by him and came under his critical analysis; and he did not spare the lawyers and the judges. But criticism was only a part of Bentham's function. He was never merely negative or destructive : his object was construction and emendation, and criticism was simply a means to this end. He aimed at being a great law reformer ; and so he set forth schemes of his own, often with much detail, showing how to rectify abuses, to supply defects, and to bring the ideal (as he conceived it) nearer to realization.

Nor were his efforts without result. It is noteworthy

that practically all the great legal reforms that Bentham advocated have been carried into effect; and one of the last of them was made only quite recently when Parliament enacted that an accused person might give his own evidence in a criminal court, without prejudice to his case. He also interested himself in, and had very definite views regarding, international law, and laid down principles of great value; and his writings on this topic served to bring the problem into prominence and to guide others in the handling of it. In all directions, he gave the lead, displaying unwonted insight and wisdom; and his place in the history of judicial reform is outstanding, and is usually acknowledged to be so. 'I do not know', says Sir Henry Maine, 'a single law reform effected since Bentham's day which cannot be traced to his influence.'

It is impossible here to give any adequate treatment of this branch of his writings. We must content ourselves with selecting several points that show the lines on which his thoughts moved.

And, first of all, he was keenly sensitive to the chaotic character of the laws of the land (largely a heritage from the past—a fact for which he did not make due allowance), and to the need of a strenuous effort at sifting them—weeding out the obsolete, discarding the useless, and classifying and explaining the remainder. This process he termed, in a word of his own construction, 'codification'. Gladly would he himself have codified the laws of the land, had he been encouraged to do so. His utilitarianism seemed to him to supply the very principle necessary for the purpose. But, failing that, he devoted himself to criticism and the exposure of confusion and absurdity, and to showing how the thing could be done, if it were honestly attempted. He put his principles into practice and formulated them for other countries (e.g. France and Russia), demonstrating in concrete instances how his theory would work; and from these examples it can be seen how thorough and how practical his procedure was.

But he looked also at laws from the side of those who were subject to them. He was strongly of opinion that knowledge of the laws of the land should be put within reach of all who are held responsible for the keeping of them. If the plea of ignorance will not save the transgressor from the penalty of his transgression (he sarcastically said that only the lawyer escaped punishment for his ignorance of the law), the State should take care that ignorance, so far as possible, be removed; which could be done by a system of general education, and by the distribution of copies of the law gratuitously or at a nominal price.

But more still is needed. If the law, which commands obedience, is to be understood by the ordinary citizen, it must be expressed in plain terms and short, easily followed sentences. Bentham was particularly sarcastic over the manufacture of the law and the extraordinary form it assumed. He criticized the drafting and the style of the laws with a vigour and a trenchancy that are refreshing still. He complained of the hideous and unnecessary technicality, of the dreary repetitions, of the redundancies, of the obsolete phraseology, of the obscurity, and demanded as plain and simple a statement as possible, in the interests of the ordinary man. He, further, gave point to his argument by adducing

telling concrete examples, culled from the laws themselves. He is always at his best when dealing with
the concrete.

Not less scathing was his criticism of the mode of
administering the law. He maintained (and with considerable justice) that obstacles almost insuperable,
pressing most hardly on the poor, were put in the way of
the litigant, or aggrieved person—unnecessary expense,
unconscionable delay, uncertainty, and vexation. For
one thing, there was no direct access allowed to the
judge. The way was barred by multifarious agents—
attorneys, barristers, &c.—heaping expense on expense
at every turn: 'In this country, justice is sold, and dearly
sold—and it is denied to him who cannot disburse the
price at which it is purchased.' In the next place, the
judge himself (Lord Eldon being the flagrant example)
deferred his verdict so long that the parties interested
were worn out with anxiety and uncertainty. Last of all
came vexation, arising from miscarriage of justice
through technicalities and the like.

But the Courts of Justice needed reform no less than
their procedure. Bentham had little respect for the
judges. He spoke of them contemptuously as 'Judge and
Co.' In what we hope is an exaggerated statement, he
wrote of them: 'As to learned judges under the existing
system, I have shown to demonstration, nor has that
demonstration ever been contested, nor will it ever be
contested, that (not to speak of malevolence and
benevolence) the most maleficent of the men whom
they consign to the gallows is, in comparison with those
by whom this disposition is made of them, not maleficent,
but beneficent.' 'Our laws', he said, 'are made by judges

for the benefit of judges,' and he was scornful of petty reforms, when the legal abuses ought to be swept away in a mass. He vehemently inveighed against Lord Eldon as the most powerful opponent of law reform; and he heartily supported juries, on the ground that they are a check on the despotism of the judges. Further, he insisted on individual responsibility in all judicial offices, and, as a corollary, advocated the propriety of only one judge to a tribunal: plurality of judges trying a case meant weakened responsibility in each; and, in the event of divided opinion, the very fact of lack of unanimity among men all supposed to be equally competent to form a judgement, had a bad effect upon the public, and gave ground for the belief, or at any rate the suspicion, that absolute justice might not have been reached after all. Then, there was the appointment of judges to their high office. That should proceed solely on merit and proper training, and partisan motives should be wholly excluded. It is no wonder that Bentham, holding these views, was very pronounced in his condemnation of allowing county gentlemen to be administrators of justice. It seemed to him to be putting a premium on ignorance and inefficiency.

The sanity of all this is obvious, and the only wonder is that it should have been reserved to Bentham to say it.

II. PUNISHMENT AND PRISONS.—As the great end of punishment is the prevention of crime, the punishment of evil-doing, in any given instance, should be exactly suited to the purpose—neither more nor less. The test, and the only unerring test, of adequacy or suitability of punishment, according to Bentham, is the good of the community, or the ability of the punishment to secure the public welfare. On this account punishment must be taken out of the hands of the partisan and of the person injured. The partisan sees justice only from the point of view of his own class or party, and, therefore, would inflict punishments that are extravagant on opponents and too lenient on friends : from the circumstances of the case, he cannot be impartial. The aggrieved person, on the other hand, brooding over his own wrong or that of his friends, and full of resentment, would administer punishment that is far too severe, because he is actuated by the feeling of self-importance or of personal partiality, and by the evil passion of retaliation and revenge. The malignity of human nature comes in, as well as the sense of self-interest and personal attachment. There is necessarily a lack of proportion between his state of mind and the punishment to be inflicted : calmness and rational consideration of the circumstances are wanting, and, while there is the absence of the judicial spirit, there is also the presence of a spirit of cruelty and vindictiveness. In either case, the consequences to the community would be disastrous.

The same regard to consequences determines the question of capital punishment for murder. This is not really a question of whether life is the inalienable property of the individual, with which no one, not even the State, has any right to interfere ; but a question as to whether in the case of murder anything short of death will serve the end in view, viz. the safety or security of society at large. Sentiment must not be allowed to overrule, nor vindictiveness, nor *a priori* theory as to abstract rights: it is solely the consideration of whether the general good demands the death of a culprit, or whether the refusal to inflict death would not do more harm to society, by encouraging others to commit crime, thereby leading to general insecurity, than the infliction of the death penalty would do evil. It is at best a choice of evils—for punishment *is* an evil; and no thorough-going or satisfactory principle can enable us to choose aright but that of the effects of capital punishment on the ultimate good of society.

In like manner, whether capital punishment should be restricted to murder, or whether it should not include other crimes (such as sheep-stealing and forgery) is determined by the same principle of consequences or the general good. If hanging for sheep-stealing has been abandoned since Bentham's day, the reason is the perception that the punishment is out of all proportion to the crime—the injustice contained in it far outweighs the justice; and if a man is not now executed for forgery, it is because it is seen that capital punishment for such an offence would be greater than the crime demands— more harm would be done than good effected by such a drastic remedy.

But what of the reformation of the criminal? That seems to be overlooked by the rigorous application of the greatest happiness principle: it might almost appear that society counts for everything and the individual for nothing. But that is not so. Society would be nothing but for the individuals that compose it; and, therefore, even the criminal members of it must be considered. And so the criminal's own good is part of the calculation

of the balance of consequences in meting out punishment. One might think that capital punishment is
absolutely antagonistic to the reformation of the criminal
himself. And, certainly, it puts an end here to further
opportunities of reformation on his part. But what if
continued opportunities should be utilized by him only
for further crime and deeper degradation, and for
contaminating society more and more? In that case,
prolongation of days would not be a boon even to the
individual criminal. But, apart from this extreme
instance, the fate of criminals and evil-doers generally
lay ever near to the thoughts and to the heart of Bentham. A large portion of his writings is devoted to
consideration of them; and many practical reforms were
advocated (including his pet scheme of Panopticon) for
ameliorating their lot and bringing them into the ranks
of the industrious and well-behaved, and preparing them
for performing their part in life as good citizens.

In order to reach a proper view of punishment and its
gradation, many things have to be considered. In the
first place, the kind or nature of the crime demands
attention. Is it heinous in its character or only comparatively trivial, i.e. does it affect few people or many,
and does it strike at vital interests or only at subordinate
ones? Is it a crime that is likely to be committed by
others, if the offender is left unpunished? In the next
place, the circumstances under which the crime was
committed matter much. Was there great provocation,
or was it deliberately and ruthlessly aggressive—planned
and committed (as the phrase is) in cold blood? What
was the doer's previous character? What his parentage?
What his physical environment and social upbringing?

Everything, indeed, of an explanatory nature must be taken into account. So too, again, regard must be paid to the motive. Was the crime a wholly selfish act, or did it involve concern for or sympathy with the condition or distress of others, as when a man steals in order to relieve his starving child? Once more, the kind of person to whom the injury has been done is an element for consideration. Was it the feeble or the helpless (e.g. children and aged or infirm people), or the able-bodied and strong that were ill-treated? Was it man or woman —directly offending or personally offensive? All these, and many more similar things, must be weighed, if justice is to be done to the offender, as well as to the community, whose interests count for most.

In reference to punishment, Bentham had a complaint against both the legislator and the administrator of justice. He accused the legislator of not paying strict regard to the grading of punishments in the laws that he enacts. Reasons for this are many, but there is one in chief, viz. that those who make the laws of the land usually belong to the higher ranks or better-off classes of society, and so estimate the effect of punishment on offenders in general by how it would affect people of their own standing. This comes out very clearly when we notice the fact of the fondness of legislators (this was specially applicable to Bentham's time) for the death-penalty as compared with imprisonment for life with hard labour. No doubt, the death-penalty appears terrific to high-born or prosperous people, for whom life is sweet and who would be weighed down by the thought of the shame and disgrace that the gallows brings upon oneself and one's family. But the ordinary criminal, with his love of adventure and the excitement of a precarious existence, puts very little value on life, and therefore, has little fear of death, while the sense of shame from an ignominious fate hardly touches him at all. On the other hand, perpetual confinement in a prison, necessitating constant hard labour which the criminal detests, is something the thought of which might certainly affect him with fear and would, if anything could, restrain him from evil courses.

Against the judge, on the other hand, the accusation brought was that he did much to frustrate the laws or to 'nullify' them by his quirks or 'decisions on ground foreign to the merits'. Hence, Bentham's unremitting onslaught on the judges. The accusation was, doubtless, in large measure relevant to the abuses of his time, but (thanks, in no small degree, to Bentham's exposure) it has little force now.

As punishment, in any given case, is intended to deter, i.e. to serve as an example to frighten others from committing the crime punished, it is necessary (so thought Bentham) to exhibit the execution of justice so far as possible to the public eye, so that intending criminals might see that the sentence is really carried into effect, and might thereby be themselves restrained. Hence, Bentham objected to such a mode of punishment as transportation of convicts or banishment to Botany Bay. He argued that sympathetic evil-doers, themselves meditating similar crime, were not thereby sufficiently impressed : on the contrary, they were thrown upon their imagination, which might very well represent the distant place of banishment as, in comparison with home, a kind of paradise, and the life there something to be envied and not dreaded. The same idea would, doubt-less, justify the exposure of the gallows to public view, as was the custom in Bentham's day.

If punishment is to act effectually as a deterrent, it must be certain and impartial in its imposition. If a culprit were punished (say, by penal servitude) for a certain crime to-day, and another, committing the same crime, were allowed to go free without punishment to-morrow, the whole effect of the law would be gone. A certain class of the community, in such a state of matters, would run the risk of the crime if there were even a chance of their being let off unscathed, just as they run the risk of detection because, as things turn out, many crimes go undetected. There must be uniformity and inevitableness about the law if the full deterrent effect of it is to be produced.

For the same reason, the power of pardoning culprits, or of remitting or diminishing the punishment fixed by the law itself, ought to be abolished. Either the legal punishment fixed is adequate or it is not. If it is adequate, then mitigation of it (even though it may assume the aspect of mercy) is an injustice and is detrimental to the deterrent power of punishment; if it is not, then it should be made so and the safety of society thereby secured. Nothing in the enforcement of punishment ought to be left to the will (which is often simply the caprice) of the individual administering it. It is obvious, however, that this should be taken along with the consideration of the reformation of the in-dividual offender, in the case of punishment extending over a series of years (e.g. penal servitude). If less than the allotted time be found sufficient to produce the end in view, then the sooner that the condemned man is set free the better, as he has now, by hypothesis, become a reformed character, and is suitable for good work for the community. This case, no doubt, would be regarded by Bentham as falling under the general principle, inasmuch as the event has simply proved (it could scarcely have been foreseen) that the fixed punishment was here too great. And, still again, remission of part of the punishment, in the circumstances, although it may not strictly speaking be deterrent, is, nevertheless, for the greater good of society, inasmuch as it shows that even criminals may look for humane treatment without strict justice being abated, and may thereby be induced to do their best towards reformation of themselves, while the general community is satisfied that everything is being done by the law to turn ill-behaved subjects into good and worthy citizens.

In connexion with punishment may be taken Bentham's handling of the prison system. Imprisonment was a penalty then attached to many more offences than it is to-day, and the treatment of prisoners was in a high degree brutal and inhuman. The prison itself, as a building, with its dismal cells and dreary dungeons, was usually a disgrace to civilization; and the way in which its inmates were fed and tended was something appalling. In particular, little attempt was made to keep the deeper-dyed criminals rigorously apart from others, or to grade treatment of culprits in accordance with the kind or degree of the crime committed. Hence such serious consequences as contamination of one by another and propagation of vice. Prisons where juvenile offenders were allowed to associate with aged

and hardened transgressors, instead of being places of
reform, became schools of vice and crime. Needless to
say, Bentham fully sympathized with Howard and his
herculean efforts at prison reform. His own great
practical contribution towards the solution of the
problem was his Panopticon, which he facetiously
described, in a letter to Brissot, as 'a mill for grinding
rogues honest, and idle men industrious'. It was really
a humane contrivance in the form of a penitentiary for
the reformation of criminals (there was an adaptation
of it also to paupers). It took the shape of a great
scheme for the erection of a peculiar type of building,
originally designed by his brother, Sir Samuel Bentham,
the distinguished engineer who achieved fame in Russia,
intended for the proper housing of criminals, for the
careful oversight of them, and for employing them in
work that would be both interesting and productive.
With his love for words of his own coining, drawn from
his knowledge of Greek, he called it 'Panopticon', to
designate the fact that the governor of it could, from his
lodge at the centre, 'see all' the occupants, whose rooms
were to be so arranged that their lives and doings should
be under constant observation. Discipline, of course,
was necessary, and careful watching of the offenders;
but it was to be tempered by sympathy and helped by
improved environment. The criminals were to be taught
to work, and not simply compelled to labour as a punishment; and, in order to create in them an interest and
ensure their coming by and by to love work, instead of
abhorring it, they were to be taught useful trades, so
that their work might be profitable and they themselves
be sharers in the profits. Thus were they to be

encouraged in the way of industry and in the acquiring of habits that would stand them in good stead against the day when they should be discharged from prison and have to face the world again. Moreover, in their leisure hours, they were to be educated to some degree on the lines of elementary education. And that nothing of an elevating and refining kind might be wanting, moral and religious training was to be brought to bear upon them: ideals were to be set before them and their sympathies enlisted in them. Of solitary confinement Bentham said: 'To think that by vacancy of mind mental improvement can be assured! It is by well filling it, not by leaving it unfilled, that I (in Panopticon) should have operated.' Every sanction that could tell was to be made as effective as possible, so that reformation might be real and stable. And, at the moment of their discharge, the criminals were to be provided with employment, until such time as they had gained or regained public confidence and were able to fend for themselves. It was a great and noble scheme—worthy of Bentham's philanthropic nature; and so enthusiastic was he over it that he was ready, without pecuniary reward, to act as its first governor. But the project was not destined to be carried out. It cost Bentham many years of labour, worry, anxiety, and disappointment—a long-continued series of efforts to win converts to his ideas among great statesmen and influential members of Parliament, and to enlist general sympathy. It cost him, also, much money; for, encouraged for a time by men in power, he went the length of purchasing a site (or sites) and running up many expenses; and so large were his disbursements that, when the whole scheme fell through, Parliament had to refund him £23,000. The reason of the collapse of the scheme was held by Bentham himself to be the opposition of King George III, whose enmity he had incurred by his active opposition to the King's policy with Denmark, in the 'Anti-Machiavel' letters.

But, though ostensibly a failure, Bentham's project had lasting good results. It attracted the attention of other countries than Great Britain, and led to partial efforts at the practical realization of it; and, even in Great Britain, Bentham may rightly claim that the vast reforms of prisons and penitentiaries that have taken place since his day, and the institution of reformatories and industrial schools, derived an impulse from him and have proceeded on the principles that he laid down.

The spirit in which Panopticon was framed may be seen in the fact that Bentham strongly objected to penitentiary *hard labour*, both name and thing. As to the name, he said: 'The policy of thus giving a bad name to industry, the parent of wealth and population, and setting it up as a scarecrow to frighten criminals with, is what I must confess I cannot enter into the spirit of. I can see no use of making it either odious or infamous. . . . To me it would seem but so much the better, if a man could be taught to love labour, instead of being taught to loathe it. Occupation, instead of the prisoner's scourge, should be called, and should be made as much as possible, a cordial to him. It is in itself sweet, in comparison with forced idleness; and the produce of it will give it a double savour. . . . Industry is a blessing; why paint it as a curse?' Then of the thing he said: 'Hard labour? Labour harder than ordinary, in a prison? Not only it has no business there, but a prison is the only place in which it is not to be had. Is it exertion that you want? Violent exertion? Reward, not punishment, is the office you must apply to. Compulsion and slavery must, in a race like this, be ever an unequal match for encouragement and liberty; and the rougher the ground, the more unequal. By what contrivance could any man be made to do in jail the work that any common coalheaver will do when at large? By what compulsion could a porter be made to carry the burthen which he would carry with pleasure for half-a-crown? He would pretend to sink under it: and how could you detect him? Perhaps he *would* sink under it—so much does the body depend upon the mind. By what threats could you make a man walk four hundred miles, as Powell did, in six days? Give up, then, the passion for penitentiary hard labour, and, among employments not unhealthy, put up with whatever is most productive.'

These are wise words, far-reaching in their application.

CHAPTER VI

JAMES MILL

THE most strenuous, and perhaps the ablest and most uncompromising, disciple that Bentham had was James Mill. Sprung himself from the people, he knew them and sympathized with them; and, true to his Scottish nature, he had the keenest interest in social and political problems. He did incomparable service to the utilitarian cause, inasmuch as he brought to bear on the Benthamite teaching his keen psychological insight and the application to it of the principles of the Associationist School, in which he was himself a master.

I. HIS LIFE.—James Mill was born on 6 April 1773 in a little village in Forfarshire, Scotland, known as North Water Bridge, on the river North Esk, in the Parish of Logie-Pert, and died in London on 23 June 1836. His father (also named James) was a shoemaker, and his mother (Isabel Fenton) a farmer's daughter. After his early education at the parish school of Logie-Pert and at Montrose Academy (where he had Joseph Hume as a school-fellow), he proceeded, in 1790, to the University of Edinburgh, where, four years later, he graduated Master of Arts, and afterwards studied Divinity, and was licensed as a Preacher of the Gospel in the Church of Scotland in 1798. His work as probationer in the ministry is practically a blank: at any rate, we know little of his exercising the gift of preaching, but find him, instead, acting as tutor in various families and pursuing historical, political, and philosophical studies. His early patron was Sir John Stuart of Fetter-cairn, who engaged him as tutor to his only daughter, and with whom, in 1802, he proceeded to London (Sir John being then Member of Parliament for Kincardine-shire), and began his literary career. In 1803, he was instrumental in starting *The Literary Journal*, which also for a brief time he edited, and to which he contributed many articles. But his power as a writer on Political Economy became first generally apparent in his pamphlet on the *Corn Trade*, in 1804. The year 1805 was the date of his marriage with Harriet Burrow. From 1806 to the end of 1817, he was engaged with the composition of his *History of India*. During this time also he wrote many articles to various journals and periodicals, e.g. the *Edinburgh Review*, the *Annual Review*, the *Philanthropist* —and, above all, to the Supplement to the fifth edition of the *Encyclopaedia Britannica*, in which appeared, *inter alia*, his well-known Essays on 'Government', 'Jurisprudence', 'Laws of Nations', and 'Education'. In 1818, he published his *History of India*—a remarkable work, which brought him immediate fame. In consequence, he received an appointment in the India House, in the department of Examiner of India Correspondence; and he became head of the Office in 1830. For the first three years of its existence, he was a regular contributor to the *Westminster Review*, which owed its origin to Bentham; and, later on, he contributed to the *London Review*, projected in 1834 by Sir William Molesworth. In 1829 was published his *Analysis of the Phenomena of the Human*

Mind; and his *Fragment on Mackintosh* appeared the year before his death, in 1835. On 23 June 1836 he died of lung complaint, in London, in his house in Kensington. His life was an extraordinary strenuous one, and the amount of work that he accomplished doubtless told upon his constitution, or predisposed him to the affection of which he died.

For many years Mill was very closely associated with Bentham, whose acquaintance he made in 1808; and he was a devoted follower to the end. Indeed, for a great part of four years (1814–17), he and his wife and family lived with Bentham at Ford Abbey. It was found, however, that their constant intercourse put too great a strain on their mutual forbearance, and, at Mill's instigation, they ceased to reside together, though continuing their intimate friendship and the relationship of master and pupil. Great though both Mill and Bentham were intellectually, they were very different in character and in temperament. Bentham's nature was essentially amiable and sympathetic; Mill's was hard and self-assertive. 'He is a character', said Bentham. 'He expects to subdue everybody by his domineering tone—to convince everybody by his positiveness. His manner of speaking is oppressive and over-bearing.' Bentham declared that Mill's political creed arose less from his love for the many than from his hatred of the few. That is a serious indictment, if true. It not simply asserts the dominance of the mind over the heart in Mill, but also indicates that Mill was deficient in the amiable virtues and was in the grasp of selfish and dissocial affection. However much this judgement may need to be modified (and considerable modification is indeed necessary), the hardness of Mill's nature, or its lack of emotion, was a fact. Even his son, John Stuart, has testified of him that 'for passionate emotions of all sorts, and for everything which has been said or written in exaltation of them, he professed the greatest contempt. He regarded them as a form of madness. "The intense" was with him a byword of scornful disapprobation.' That he was a supremely interesting conversationalist of the didactic type, is attested on all hands by associates and intimate friends (such as George Grote); and his son, J. S. Mill, has said the same thing.

II. PSYCHOLOGY.—Bentham's philosophy was weakest on the side of psychology. Not that he was unable to analyse mental states or to estimate the character of springs of action: quite the contrary. But his interests were too limited and his outlook on life, owing to his hermit mode of living, too narrow to allow an adequate handling of human nature, which, whatever else it may or may not be, is wide and complex and varied. It is Mill's distinction that he supplied a thoroughgoing psychology to utilitarianism. This he did in his *Analysis of the Phenomena of the Human Mind*, which is a landmark in the history of mental science; supplemented by his *Fragment on Mackintosh*—critical to a degree, but also expository of Mill's own system.

Mill had been educated at Edinburgh University, and so came under the influence of Dugald Stewart and of the Scottish philosophy. Of Stewart as a cultured and inspiring lecturer he spoke in the highest terms of praise to the very end of his life. His own treatment of psychology, then, must needs bear on it to some extent the Scottish stamp. It looked at the subject from the standpoint of experience and of common sense, and it developed the theme on the lines that the Scottish School, led by Thomas Reid, had made familiar. The meaning of this is that Mill's psychology is in large measure analytic and descriptive. It deals with psychical facts and processes as experienced—the nature of the mind as we find it to be, the phenomena of consciousness, the laws that these phenomena obey, and such like; and it deduces its theories from the facts, and does not force the facts into the mould of preconceived theory. Coupled with this are exceptional clearness of thought and lucidity of expression. Mill felt (as Bentham too had done) that error lies in mental fog, and that the safeguard is clear concepts and accurate definitions, just as Socrates long before had declared. A very striking feature of Mill's psychology is his persistent recourse to definition, and his dogged insistence on discriminating shades of meaning in psychological terms. In this way, he very properly tried to rid psychology of the opprobrium of a popular and unscientific use of language, or a lazy and contented employment of ill-defined words. He had also the power of exposing philosophical pretentiousness and partisanship by vigorous invective, as is seen in his *Fragment on Mackintosh*.

The psychology of Mill is a wide subject, and we must contract it here. Let us, then, ask as relevant questions : What is Mill's psychological method? How does he treat association? What is the place of pleasure and of pain in his system? The answer to the last of these questions involves his ethical theory.

Mill's method is inductive and experiential—the

vigorous application of introspection or inner observation to the study of mental phenomena. He lived too early in the day for an appeal to experimental psychology, or for anything more than an occasional reference to animal psychology; it would be unreasonable to expect him to be fully appreciative of the valuable aid to introspection that is to be found in the comparative method of investigation: psycho-physics, child psychology, social psychology as yet were not. But, on the other hand, there was physical science with its luminous suggestions, and, in especial, chemistry had come to the front; and the close relation of mind to body was becoming more and more obvious through the advances that were being made by physiology.

Indeed, it was the procedure of science that then guided the empirical psychologist. Just as, in science, the atomic theory seemed to give a full and satisfactory explanation of matter, so, Mill thought, the mind itself might be fully explained by conceiving it as constituted of sense atoms, combining and working in definite ways and under definite laws—ways and laws that could be accurately determined and scientifically formulated. Hence his doctrine of association, regarded as 'mental chemistry'. The explanation of the mind to him was just the exposition of the mode of combination and coalescence among its varied states, and what transformations this process could effect. Had he lived to-day, when the *biological* conception of mind holds the field— mind conceived as an active, living, organic unity—he would have found that there are difficulties in this conception of the mind as an organism that his theory cannot surmount.

To Mill, the one law of association is that of contiguity: the other commonly-accepted law, viz.
similarity, he is disposed to resolve into contiguity, or
frequency of conjunction. His resolution is of very
doubtful validity. His expression of the law is this: 'Our
ideas spring up or exist, in the order in which the
sensations existed, of which they are the copies.' The
order may be of two kinds—synchronous or successive.
The causes of strength in association are resolved by
him into two—the vividness of the associated feelings,
and the frequency of the association. He fails to grasp
the full significance of 'interest'—the basis of attention.

'Frequency' plays a very important part in his
associative psychology. It explains 'inseparable association'—such as we find in our belief in an external
material world, or in our inability to think colour
without also thinking extension. The value of frequency
of conjunction in explanation of knowledge is undoubted,
but it will not bear the full strain placed upon it by Mill.
In such a case as our inability to dissociate extension
from colour in our thoughts, there is another feasible
explanation—namely, that it arises from the fact that
the eye, through which we see and obtain colour, is
itself an extended organ, powerful in its muscular
movement. Nor can inseparable association adequately
explain belief. Of Mill's doctrine of belief, Bain (another
leader of the Associationist School) says: 'When James
Mill represented belief as the offspring of inseparable
association, he put the stress upon the wrong point. If
two things have been incessantly conjoined in our
experience, they are inseparably associated, and we
believe that the one will be followed by the other; but

the inseparable association follows the number of repetitions, the belief follows the absence of contradiction. We have a stronger mental association between "Diana of the Ephesians" and the epithet "great" than probably existed in the minds of Diana's own worshippers; yet they believed in the assertion, and we do not.' In Bain's opinion, as we shall see later on, belief must be based in primitive credulity and the absence of contradiction.

How, according to Mill, we come to have a knowledge of an external world is by building it up for ourselves through the sensations of the various senses associated in our experience, e.g. touch with sight and locomotive activity. The book that I now look at appears as it does to me—solid, extended, shaped, external, &c.— because the sensations of colour that my eye gives are associated in my experience with the hardness of the book to my touch and its resistance to my muscular energy. This is the associationist explanation, which has been prominent in psychology since Mill's day: an external object is an ideal construction, not an original datum of immediate consciousness.

By association Mill explains the nature and working of memory, imagination, conception, and every process and 'faculty' of the mind. But by the same principle he explains also our ethical nature. Conscience and our moral feelings and affections are to him not simple and original, but complex and derived; and the mode of their production may be traced. The ultimate elements are pleasurable and painful sensations; and how these are manipulated by experience so as to form the final product is the problem.

Take as an example the highest of the moral feelings, benevolence. This seems to be entirely outside the region of self-interest. And yet Mill solves the difficulty by insisting (as Bentham also had done) on the distinction between motive and intention. In benevolence, the motive, according to him, is always self-interest—the pleasure that the individual derives from the benevolent action; but the intention is disinterested. In other words, a benevolent action ministers to the individual's own pleasure, even though what he has in view is the good or happiness of others. Wherefore, Mill pointedly asks, 'Can any greater degree of social love be required than that the good of others should cause us pleasure; in other words, that their good should be ours?' This being so, we can easily see why he should regard the phrase 'disinterested motives' as a contradiction in terms, and the dispute about the disinterestedness of human nature as a mere war of words.

That which makes the distinction between moral and immoral acts is utility; and the means whereby we create in a man an interest in the doing of acts that are useful (and to which he is not inclined) and the non-perform-ance of the opposite kind of acts is a certain distribution of the good and evil that we have at our command. When this distribution is such as can be applied by the community in its conjunct capacity, we call it law; when it is applied only by individuals in their individual capacity, we know it as the control of the moral senti-ments.

Pleasure and pain, then, are of the essence of morality; and moral approbation and disapprobation, praise and blame, are applicable to man as a rational being who has the power of appreciating the consequences of actions in the light of an addition to or a diminution of the sum pleasures.

Mill's associationism, carried throughout all the region of psychology, set the example for his school.

III. THEORY OF EDUCATION.—Nearly related to his psychology is Mill's theory of education. He was not less alive to the value of education than Bentham had been, and was equally insistent on the necessity of educating the lower classes as well as the higher. He took a very active part in the controversy over the Lancasterian and Bell systems, and by his pen, in the *Philanthropist* and elsewhere, vigorously and strongly opposed the introduction of Church of England doctrinal teaching into the schools. He went even farther and took up the purely secularist position that religion should not be taught in the schools at all, and, therefore, that the Bible should be excluded. How keen his practical interest in education was is further seen (by the prominent part that he took in the attempt which ultimately failed) to establish a Chrestomathic School on the principles that we have already found laid down by Bentham, and by his activity (crowned this time with success), along with a few other noted educational enthusiasts, in the originating University of London

But, besides this, Mill has a specially prominent place among educational writers, inasmuch as he expounded the theory of education with a rare breadth of view on purely philosophical grounds. This is seen best in his article on 'Education' in the *Encyclopaedia Britannica*, where his comprehensiveness of outlook and his psychological insight are outstanding features. Three of his leading points may here be adverted to.

In the first place, he conceives the end of education to be—rendering the individual, as much as possible, an instrument of happiness, first to himself and next to other beings. That gives the definition of education as 'the best employment of all the means which can be made use of, by man, for rendering the human mind to the greatest possible degree the cause of human happiness. Everything, therefore, which operates, from the first germ of existence to the final extinction of life, in such a manner as to affect those qualities of the mind on which happiness in any degree depends, comes within the scope of the present inquiry.' This is clearly a very wide view of the subject to take—very proper, but very unconventional. It obviously involves both the intellectual and the moral culture of the individual; the moulding of his character, as well as the developing of his intelligence and the increasing of his knowledge, not only during his 'school' days, but throughout his whole life. Indeed, the social value of education is emphasized by Mill (as behoved a good utilitarian), and he returns to it again and again. The individual has to be trained as much in justice and benevolence as in intellectual faculty and command over the materials of knowledge : he is not an individual solely, but a social individual. Temperance or self-control is to be a chief virtue with him : he must learn to restrain his desires, and, as far as possible, to assimilate his own pleasures and pains to those of his fellow men.

For this end, Mill bases the science of education on

psychology, and elaborates the importance of the associative processes for effective educative work. Through Bain and others, this has now become a commonplace in education; but it was not so then. 'It is', he says, 'upon a knowledge of the sequences which take in the human feelings or thoughts that the structure of education must be reared. . . . As the happiness, which is the end of education, depends upon the actions of the individual, and as all the actions of man are produced by his feelings or thoughts, the business of education is, to make certain feelings or thoughts take place instead of others. The business of education, then, is to work upon the mental successions.' That is very sound teaching. Education has also to do with the influence of the body on the mind; and Mill, following Cabanis, comes to the very modern conclusion that 'an improved medicine is no trifling branch of the art and science of education'. Nourish the body, if the mind is to be vigorous. The full force of this we are duly realizing to-day, and following up the precept by practical effort. 'The physical causes must go along with the moral; and nature herself forbids, that you shall make a wise and virtuous people, out of a starving one. . . . This or that individual may be an extraordinary individual, and exhibit mental excellence in the midst of wretchedness; but a wretched and excellent people never yet has been seen on the face of the earth.'

The power of education seemed enormous to Mill, and he held extreme views on the extent of it. He followed Helvetius, and followed him too literally. His enthusiasm forgot the necessary qualifications, and he practically maintained that education is all-powerful—wholly so with regard to classes of men, at any rate, where the differences, he thinks, are undoubtedly due to education. 'If education does not perform everything, there is hardly anything which it does not perform.' 'This much, therefore, may be affirmed on the side of Helvetius, that a prodigious difference is produced by education; while, on the other hand, it is rather assumed than proved, that any difference exists, but that which difference of education creates.' Helvetius had laid it down that the mass of men are by nature equally susceptible of mental excellence, and that they differ only through education. This is substantially Mill's doctrine too. There is clearly a great truth in it, if we interpret education in the wide sense that Mill did, including in it environment in its various aspects—physical and social, and operating all throughout one's life. Difference in opportunities and surroundings, in atmosphere and experiences, does largely explain the difference that we find among men. But if we put the doctrine in the unqualified form—as, for instance, Sir William Jones did, nearly expressing Mill's view—'that all men are born with an equal capacity of improvement'—that is a very disputable position. It ignores the fact that there is such a thing as native aptitude and degrees of it, owing to heredity; that there are both individual and racial differences among men by nature; and that the power of education is limited in different cases by the material that it has to work upon. Yet, the power of education is undoubtedly vast; and we are much more likely to achieve great things in the way of mental improvement if we keep steadily before us its potency and virtue than if we be constantly harping on the natural differences that exist among men in mental capacity, and the ineffectiveness of education to make all men intellectually equal. Mill was too optimistic, but he looked in the right direction.

CHAPTER VII

JAMES MILL AS POLITICIAN AND AS JURIST

THEORY OF GOVERNMENT; POLITICAL ECONOMY; JURISPRUDENCE AND INTERNATIONAL LAW

I. THEORY OF GOVERNMENT.—Starting from the conception that the end of government is the public good, or the greatest happiness of the greatest number, it follows that the science of government must rest on the science of human nature. What experience shows with regard to human nature (so Mill taught) is, that the individual is continually prompted in his actions and his purposes by the desire for pleasure, and that he aims at securing as much of it as he can, avoiding pain as far as possible. In this pursuit of pleasure or of happiness, he is ready, if need be, to infringe on the happiness and pleasure of others—to grasp at whatever he desires or thinks would minister to his own satisfaction, in disregard of the interest or the desire of his fellows. It is a law of human nature that if a man be given power over another, he will abuse it, or turn it to his own selfish ends: in other words, he will aim at making his fellows his instruments, and will set himself, by whatever means are at his command, to subject them to his will. Hence the need of some restraining force, which shall curb and limit his desires and prevent his infringement on the desires and rights of others. It is from this that the idea of government arises; for government is just the protection of a person against the encroachment of others.

But a government is itself a body of individuals, with all the passions and tendencies of human beings in them. If unrestrained, a governing body (be its members few or many) will make its own interest supreme, will work for its own ends and benefit, and, therefore, will tyrannize over, or deal unjustly with, those who are subjected to it, ruling by terror. The graspingness of human nature is itself illimitable, and it is manifested in every association of men as in every individual.

What security, then, can we have against the abuse of power on the part of the governing body? That is the supreme question in determining the ideal or best system of government.

If we look at the three commonly recognized simple forms of government (the democratic, the aristocratic and the monarchic), we soon find that the desiderated security is not given in any one of these. Self-interest comes in, in each of them, to vitiate its working. Nor is the difficulty surmounted in the union of the three forms in what is known as the Balance of the British Constitution (King, Lords, and Commons). Such a balance is only fictitious; for any two of the constituent members may combine (and, in certain circumstances, will combine) against the third and render it impotent; and the natural combination, on the score of mutual interest, is between monarchy and aristocracy against the commons.

Is there, then, a possible security, and, if so, where does it reside? Mill was ready with his answer: A real security is possible, and it lies in the representative system—in government by the people's representatives acting as a check on legislative abuse. But, in order that they may be a thoroughly effective checking body, they must both possess sufficient checking power and also have their interest identical with that of the community —otherwise, they will make a mischievous use of their powers.

The first requisite clearly demands that the House of Commons, which is theoretically the checking body, shall be powerful enough, in a case of conflict, to counterbalance both Lords and King. Mill had no inveterate objection to a king. On the contrary, in his article on 'Aristocracy', in the *London Review* of January 1836, he laid it down that the interests of the king are bound up in the interests of the people; and that it is only when the king submits himself to the aristocracy and puts himself in opposition to the people that he becomes a curse. Yet, he may do otherwise; and then it shall be well with him.

His mode of dealing with the House of Lords is more drastic. 'Let it be enacted', he proposes, 'that, if a bill, which has been passed by the House of Commons, and thrown out by the House of Lords, is renewed in the House of Commons in the next session of Parliament, and passed, but again thrown out by the House of Lords, it shall, if passed a third time in the House of Commons, be law, without being again sent to the Lords.' This is interesting in view of the present position of the House of Lords question.

As to the second of the requisites, the difficulty is,— How can we secure that the people's representatives shall continue to identify their interest with that of the community? Mill's solution is very definite,—Limit the duration of the power of the representatives. This means frequent appeal to the electors; which, although some may think it impracticable, is not an impossible thing, and may even be made comparatively easy.

This settled, there next comes up the question of the qualifications of an elector. Here Mill expressed, in some respects, peculiar views. He maintained that the suffrage should not extend—or, on his principles, need not extend—to the members of the community whose interests are already secured in those of others with whom they are immediately associated. This at once struck off children, who are dependent on their parents, and women, who, as daughters or as wives, are identified in interest with their fathers or their husbands. There remained only men; and the point was to determine the age at which an electoral vote might reasonably be claimed by them. Mill threw out the suggestion of fixing it at forty, on the ground that men of forty have a deep interest in the welfare of the younger men, and that the majority of older men have sons, whose interest they regard as an essential part of their own. This, he said, 'is a law of human nature'.

But to the lowering of the franchise, so as to bring it near to *universal* suffrage, even universal *male* suffrage, there was a strong objection on the part of many in Mill's day. It was placing the power in the hands of the people, they said, and the people were incapable of acting in conformity with their own interests. This distrust of the people was regarded by Mill as proceeding from a 'sinister' interest; he thought that it expressed the selfish dread of the wealthy and the aristocratic ranks of society. The accusation was, moreover, unfounded; for the people could quite well recognize and appreciate their interests, although they sometimes made mistakes. They depended largely for guidance on the counsel and wisdom of those better instructed and higher in the social scale than themselves; and, through the dissemination of knowledge, they would more and more realize their true interests, for it is not possible for a community really to know what is for its good and yet persistently to act against it—to know what is conducive to its happiness and yet prefer misery. To the middle rank of society Mill looked for a solution. The middle rank was the section of society 'which gives to science, to art, and to legislation itself, their most distinguished ornaments, and is the chief source of all that has exalted and refined human nature'; it was, besides, the section of society in most immediate and sympathetic contact with the lower classes, whose intelligence and virtues they respected, and the members of which they were wont to look up to as models for their imitation. Therefore would this middle rank be naturally the class to influence and guide the people. This pronouncement is very interesting, if somewhat surprising: it shows, at any rate, Mill's own estimate of the grade in society to which he and his fellow utilitarians belonged, and the claims that they made to be the natural guides and counsellors of the people.

Such is the substance, in brief, of Mill's theory of government, which was the grand authoritative presentment of the utilitarian views at the time. That it has vulnerable points is evident. For one thing, it starts with a particular conception of human nature, and therefrom deduces the whole theory of government, apparently apart from appeal to experience. To this it may be objected that Mill's analysis of human nature is partial, being simply the recognition of self-interest in man, without any adequate acknowledgement of the important part played by man's social nature in his relation to his fellows—by fellow-feeling, sympathy, generosity, and the like. Further, to Mill's deductive method objection might be taken on the ground that it is not from assumed principles, or by abstract reasoning, that the nature of government is to be demonstrated, but from experience of what is actual, carefully studied, with generalizations made on the inductive or Baconian method. This was the line of attack that Macaulay pursued in his vigorous and brilliant articles in the *Edinburgh Review*. But, again, Mill's teaching might be attacked on the side of its theory of representation. If the guiding principle be the identity of the interests of the representatives of the community with those of the community itself, that means democratic tyranny, or an abuse of power by the 'masses' ignoring the interests of the 'classes', not less flagrant than that of the aristocracy against the community, for which the theory was brought forward as providing a sufficient check. This was the line of argument adopted by Sir James Mackintosh, and constraining him to advocate a scheme of parliamentary reform grounded on the representation of classes.

But, be all this as it may, the point is that Mill's theory had far-reaching and lasting consequences in British politics, appealing to practical politicians of the time and captivating the masses. The *Encyclopaedia* article on 'Government', in which the doctrine is most fully set forth, was no mean factor 'in the train of events culminating in the Reform Bill of 1832'. That article and allied writings had enormous influence: they were scattered far and wide throughout the land. Every enthusiastic supporter also became a propagandist. A striking instance lies to hand. When Joseph Hume was Rector of Marischal College and University, Aberdeen (1824, 1825, and 1828), he presented a copy of the reprint of Mill's *Encyclopaedia* Essays (a book stated on the title page to be 'not for sale'), inscribed 'With Mr. Hume's compliments', *not* to the University Library, which one would naturally suppose to be the proper recipient of such a gift, *but* to the Library of the Mechanics' Institution. Thus, by going direct to the people, could Mill's teaching be made to reach the intelligent artisan and the aspiring youth of the lower classes, who were zealously pursuing self-culture and were eager for light at the moment when things political were gathering themselves up for a great issue.

But it was not only by his writings that Mill stimulated and guided political opinion; he worked effectively through his personal influence on great politicians in Parliament. His authority was potent with Lord Brougham; it was supreme with George Grote; Ricardo was his disciple, in all save political economy, where Mill followed; Joseph Hume was his strenuous ally; he captivated Roebuck, who produced cheap issues for the people of many of his articles. In every way, he was the leader of the Utilitarian Radicals, after Bentham, and the chief operative force in effecting the practical reforms of the school.

II. POLITICAL ECONOMY.—To Mill, as to Bentham, Adam Smith was of first importance in political economy; but, in Mill's time, two other forces had come into play, and he fell under the influence of both. One was T. R. Malthus (1766–1834), who had awakened the thinking world by his speculations on Population; and the other was David Ricardo (1772–1823), whose economic doctrines attracted wide attention among economists.

The great end that Malthus had in view was that of the Utilitarians in general—human happiness or the improvement of society. To further this purpose, he aimed, in his *Essay on the Principle of Population* (first published in 1798), at discovering the conditions and possibility of progress, with special reference to the relation between population and means of subsistence. The vision of the Perfectibility of the race had risen before the eyes of many thinkers, French and English, and was specially associated at the moment with Condorcet and Godwin. To Malthus perfectibility seemed impossible so long as population tended to outrun subsistence. Accordingly, he set himself to a thorough investigation of the subject; one result of which was the disclosure of the fact of the appalling relation between the growth of population and the means of subsistence. Malthus expressed it in the proposition that, whereas population, if unchecked, increases in geometrical progression, the means of subsistence increases only in arithmetical progression. The question with him, then, came to be: How is the rapid increase of population over means of subsistence, within a definite area, to be stopped? His answer was: Through 'checks'. These he reduced ultimately to three, viz., misery, vice, and moral restraint. Nature herself provides checks, apart from human foresight or interference. But foresight and rational interference may do much. And this is the line along which the problem of population seemed possible of solution. Two points, with a very practical bearing, may here be specially mentioned. One is, the institution of private property. By this means, Malthus saw, people would be stimulated to industry and prudence. The other is, correct views about Marriage. This involves arousing in people a due sense of the responsibility that marriage entails—responsibility for possession of the means of living necessary for maintaining a household before marriage is entered upon, and responsibility regarding increasing the population and the need of moral restraint.

This Malthusian doctrine was fully endorsed by Mill, and became part of his political creed.

Not, however, in general economic theory did Malthus affect Mill. The potent influence here was Ricardo—one of Mill's most intimate and cherished friends. It was Mill who induced Ricardo to write and to publish his notable work on *The Principles of Political Economy and Taxation* (1817); and it was he, too, who successfully instigated Ricardo to enter Parliament in 1819. On the other hand, Mill's *Elements of Political Economy* (1821) is thoroughly Ricardian in doctrine and in spirit. He himself is very modest in his claims regarding it. He does not profess to be introducing original notions, but to be simply presenting accredited doctrines in such form that they could be easily understood by any person of ordinary intelligence willing to give attention. 'I cannot

fear an imputation of plagiarism', he writes, 'because
I profess to have made no discovery; and [in excuse for
not quoting authorities] those men who have contributed to the progress of the science need no testimony
of mine to establish their fame.' The presentation itself,
however, is extremely effective. The economic problems
are set forth with admirable clearness and precision of
statement, and in a logical order that leaves nothing to
be desired. If the treatise is 'a school-book' (as he
designated it), it is a particularly good one, and served
well the purpose of making the subject of political
economy, from the Ricardian point of view, accessible
to the general reader.

III. JURISPRUDENCE AND INTERNATIONAL LAW.—The
enthusiasm for law and law reform that characterized
Bentham was shared by Mill. Mill had great ideas on
the subject, and contemplated a work on Conveyancing
and another on the History of English Law, and, later
on, a full exposition of a system of Jurisprudence. These
were largely what he calls 'projects which float in my
head'; but the last of them was carried out to an appreci-
able extent in his two authoritative articles on 'Jurisprudence' and the 'Law of Nations' in the *Encyclopaedia
Britannica*.

If we take the statement with the necessary qualifications, jurisprudence may be said to deal with rights. The
necessary qualifications are, that it is not concerned
with the *creation* of rights—with the question as to what
ought and what ought not to be rights; nor with the
question of the distribution of rights—of how rights can

best be distributed so as to produce the greatest general happiness. These questions belong to legislation. What jurisprudence has for its province is the protection or security of rights. In a passage in the *Fragment on Mackintosh*, Mill puts it very lucidly, thus: 'Rights, jurisprudence takes as it finds them; and then inquires by what means they can best be secured. By its investigations it has established that for this security it is necessary, first, that rights should be accurately defined; secondly, that such acts as would impair or destroy them should be prevented by punishment; thirdly, that men should be appointed to determine all questions relating to rights, and the violation of them; fourthly, that the trust vested in each and the mode of exercising it should be according to certain principles, and fixed by rules. Definition of rights, punishment for wrongs, constitution of tribunals, mode of procedure in the tribunals, are the heads under which all the objects of Jurisprudence are arranged.'

As to the two branches of law, civil and criminal, it is the province of the Civil Code to define rights and of the Penal Code to define and characterize offences and punishments. On the other hand, the nature and constitution of the courts of justice, operating by judges and dealing with disputed claims and evidence, is the subject-matter of the Code of Procedure. These various topics Mill takes up and elaborates in his article on jurisprudence. The treatment is fresh and vigorous, but does not, to any large extent, advance beyond Bentham.

It is different when we come to international law. Here Mill makes a very distinct and independent contribution.

It is obvious that, in its application to nations, the term law has a different signification from what attaches to it when applied to the individuals constituting a single people or community. In the latter case, law pre- supposes a properly constituted authority (the Govern- ment) issuing commands and enjoining obedience (which, on the hypothesis, it has a right to do, and which people are wont to acquiesce in), with the power of enforcing the commands, if need be, through the applica- tion of pains or penalty—punishment for disobedience. But, when different nations and their relations to each other and their conflicting interests are under considera- tion, the case assumes another aspect. There is no universally-recognized central authority here (so it was in Mill's day) which rival countries approach and to which they spontaneously submit; nor would it be easy, Mill thinks, to get the nations to combine to form such, for 'nations hardly ever combine without quarrelling'. And if there is no such generally-accepted authority, with the power of final decision, there is, of course, no common reserve of punitive power, to which appeal might be made to get a recalcitrant nation to desist from active hostility. If there is a sanction that applies to nations at all, it is clearly not of the same nature as that which enforces obedience in the case of the laws of the land. Of what nature, then, is it? It is the *popular* sanction—public sentiment, or the force of general opinion brought to bear on a particular commonwealth, constraining, but not forcibly coercing, it to fall in with or accept the course of action or line of policy that commends itself to the others or, at any rate, to those of them that are civilized. This, like the ordinary operation of public opinion on the individual, appealing to his humanity, moulding his character and affecting his conduct, if properly expressed, may be exceedingly strong and effective. Not even a powerful country can light-heartedly pursue a policy that goes in direct opposition to the unfavourable sentiments of the rest of the civilized world; least of all if that country be democratic in its constitution. This last qualification is specially to be noted.

The maxims and rules, then, that we know as international laws, although peculiarly sanctioned, are, according to Mill, very far from being worthless. They bind after the manner of a code of honour among gentlemen, whose effective working is dependent on the general opinion of the society affected by them.

For the full and satisfactory working of these international rules, two things are requisite; (*a*) a code of laws, and (*b*) a tribunal for the administration of them, with a well-defined mode of procedure. Are these things possible? Mill gave a very definite answer in the affirmative.

A code of international laws means simply the determination of the rights of nations; and these, he held, may in large measure be formulated. For example, a nation has rights in time of peace to its own territory and its watercourses or rivers, and to a share in the open sea for commerce. Oceanic communication of one country with another must be allowed to every country: each, thus far, has an equal right. Rights of nations are created just as rights of individuals are—by such things as original occupancy, transfer through contract, conquest; and a nation's rights are brought to an end by causes similar to those that terminate the rights of the individual—willing transfer, &c.

But what of a nation's rights in time of war? That is ever a pressing question. Mill was very strongly convinced that all nations gain by the free operations of commerce, and so he advocated free traffic all round, so far as concerns the property of individuals in time of war; and he strongly condemned piracy. In this way, he thought, an end would be put to the difficulties and disputes about the maritime traffic of neutrals. He brought out clearly that what justifies entering on a war also determines when a war ought to cease. If the legitimate object of a war is compensation for an injury received and security against future injury, then a successful war ought to terminate immediately on the attainment of that object.

Mill's whole handling of the subject of war, and of the rights and obligations of belligerent nations, is masterly. But it is necessarily limited by the circumstances of his time, and needs to be further developed and modified to-day. What when a war is undertaken from national ambition and with a view to selfish dominance? The situation then needs further consideration.

Nor had Mill much difficulty about the establishment of a tribunal for the administration of international law, and an effective mode of procedure. The desirability of such a tribunal is generally granted. It would not only be a solvent of present difficulties, but would also operate as a great school of political morality. The difficulty is: How could nations be got to submit to it? That is a question that confronts us to-day, as much as it did Mill.

Mill answered: Through the sanction of general opinion. Given a properly constituted tribunal, duly representative of the nations, dealing impartially with the cases brought before it for decision, and given the decisions and proceedings of the tribunal made publicly known and promulgated throughout all the countries of the civilized world, then the general utility of such a body would very readily be seen and its power felt. It would soon be discovered that many kinds of international disputes would be more satisfactorily determined by an appeal to the tribunal than by the hot-headed arbitrament of the sword. Indeed, there might by and by be created in the nations so strong a public feeling that differences between rival or conflicting countries should be settled by arbitration, that refusal to appeal to the international tribunal would be taken practically as a confession that both parties to the quarrel were in the wrong. That is true; but how does it avail in the case of a nation or nations contemptuous of public opinion? Even then, he thought, a decision by the tribunal, although it would not affect the contemptuous parties, need not be utterly useless. It would be a benefit to the other nations. 'If these decisions constitute a security against injustice from one another to the general community of nations, that security must not be allowed to be impaired by the refractory conduct of those who dread an investigation of their conduct. . . . A decision solemnly pronounced by such a tribunal would always have a strong effect upon the imagination of men. It would fix and concentrate the disapprobation of mankind.' For the creation of a moral sentiment that would by and by act as a strong restraining power on the injustice of nations, Mill trusted also to making the book of the laws of nations, and selections from the book of the trials before the international tribunal, a regular subject of study in every school and a knowledge of them a necessary part of every man's education.

Mill had a vision of the nations of the world at amity, each subordinating its own interests to the interests of the whole, and, therefore, each content to mind its own concerns without unduly interfering with the concerns of its neighbours or wishing to lay hold of its neighbour's territory. The principle of utilitarianism was supreme with him, and he necessarily deprecated anything national that would be of a selfish or individualistic character, anything that would be incompatible with the interests of the nations in general, or, at least, of the civilized nations, which presumably, in the long run, means that of the uncivilized nations too. There were trends in the nineteenth century which seemed to promise the realization of this vision. The scientific discoveries of the century, with their unprecedented results in application to man's practical needs—as seen in railways, in the telegraph, in navigation, and the like —practically annihilating space and time, and bringing peoples all over the face of the earth into immediate contact (commercially, politically, intellectually, and socially) in a way and to an extent unheard of before, led to the general diffusion of humane feeling and kindly sentiment, and, therefore, to a better acquaintance of one people with another, and a mutual sympathy that gave promise of lastingness. The trend was continued during the first years of the present century. In the civilized world, there seemed to be a readiness and even a genuine desire to submit international differences or rival claims to arbitration, if not to prevent the unreasonable pushing of rival claims altogether. The utilitarian principle appeared to be effectively at work in its highest form; the idea of the general good of mankind laying hold of the nations, as well as of individuals, and operating for universal concert. Hence the Hague Convention and the Palace of Peace—so full of hope. As late as 1913, men rejoiced in the harmonious settlement of the Balkan War through the great Powers working together as one community to preserve the peace of Europe; and Lord Haldane, speaking as Chancellor of Great Britain, in his brilliant address to the American Bar Association at Montreal on 1 September of that year, could set forth with confidence his group-system scheme, which was based on belief in the potency of cherished good-feeling between members of the same group of kindred nations—say, the Anglo-Saxon or English-speaking peoples of the United States of America, Canada, and Great Britain—constraining one member of the group to pay regard to and take a sympathetic interest in the point of view of the others and disclosing a common ideal that would keep the whole group in unison—a force working, not legally but ethically, yet powerfully, like the general will of a single people at a great national crisis, which sweeps everything before it.

But the great set-back has come with the present disastrous and barbarous European war, which has given an unspeakable shock to every lover of peace. If the teaching of General Friedrich von Bernhardi and Treitschke is to be taken as voicing German opinion, then the position is—(1) That a nation is not bound by the same morality that is incumbent on the individual— something happens, when citizens join themselves together as a people and cultivate the spirit of patriotism or fatherland that absolves them from all the fundamental ethical principles that bind one man to another and neighbour to neighbour; and (2) that a nation's word of honour, its treaties and its contracts, have no lasting binding force, but are simply 'scraps of paper', to be torn up and thrown aside whenever the nation finds it to its advantage to do so. The only ruling principle is 'Might is right'. This Machiavellian morality is surely abhorrent to the civilized conscience of the twentieth century. The situation shows, however, that neither public opinion nor friendship and goodwill, however sedulously cultivated between one nation and another, is sufficient to secure general amity when a nation's selfish ambition comes in and overrides every other principle; but that an international tribunal must have behind it the sanction of physical force—the means of actually enforcing its will—if it is to be the power for good that peace lovers desire. One can only hope that, in process of time, through the alchemy of the emotions, there may emerge from this such a feeling of the solidarity of the race as shall be able to dispense with the spectre of superior physical force lying behind, and leave the sense of brotherhood and unity among men the sole constraining and efficient power.

CHAPTER VIII

JOHN STUART MILL

I. LIFE AND WRITINGS.—Of the children of James Mill, John Stuart Mill was the eldest. He was born in London on 20 May 1806 and died at Avignon, in France, on 8 May 1873. The unique system of education to which he was subjected in his early years by his father is so well known that only a brief reference to it need here be made. Set to learn Greek at the age of three years and kept steadily at the task till he attained the age of eight, with English and arithmetic brought in only in a subsidiary way, he became early enamoured of Greek thought, and, in particular, of the dialogues and dialectic method of Plato. At the age of eight, Latin was added to Greek as a subject of classical study; and, before he reached the stage of adolescence, he was by degrees introduced to the higher disciplines of logic, psychology, and political economy. These subjects were 'stiff' subjects, usually reserved for men of maturer years. Up to this point, his father was his sole teacher, careful and exacting, methodical and thorough, with strong personal convictions as to what to teach and how to teach it. He made the boy his constant and intimate companion, sharer from the earliest possible moment of his daily walks and talk, as well as pupil in the study. In this way, he not only exercised a watchful supervision over his work and attached him to himself, but also, by a process of Socratic cross-questioning and didactic discourse, gradually evoked and trained the powers of his mind within a few years to an extent that is marvellous and probably unprecedented. He also set him early in life to act as monitor to the younger members of the family, and thereby furthered his intellectual development by making him feel that teaching a subject to others is the best way of understanding it. The consequence is that the precocity of young Mill has become proverbial.

When the boy reached the age of fourteen, a new turn was given to his life by his being sent for a full year to France, as the guest of Sir Samuel Bentham (the brother of Jeremy), living with him at Toulouse and at Montpellier, and making excursions to the Pyrenees and elsewhere, much to his delight. In this way, parts of his nature were touched that had hitherto lain mostly dormant. Not only did he acquire a knowledge and a facility in the use of the French language, and make himself acquainted with French literature and politics, thereby contracting leanings and likings that were to influence him greatly in the future, but he became a passionate lover of nature and a zealous student of botany and zoology, and developed a love for travel, which continued with him throughout his life.

A fresh stimulus was given to his thinking, soon after his return to England, by his gaining access through his father to the *Traités de Législation* of Dumont, which was Bentham's ethical and political speculations clothed in French garb and expounded by a distinguished Frenchman. 'The reading of this book', he says, 'was an epoch in my life; one of the turning-points in my mental history.' Coincident with this was another helpful factor in his mental training—his studies in Roman law, under the congenial guidance of John Austin, the Jurist.

At the age of sixteen, he conceived the plan of a small society of young men like-minded with himself, for the discussion of ethics and politics on Benthamite principles, to which he gave the name of 'The Utilitarian Society'. It was duly formed, and continued in existence for three years and a half. A little later, he joined 'The Speculative Debating Society'; and was also a prominent member of a youthful band of thinkers who met for discussions in George Grote's house, the subjects discussed being political economy, logic, and psychology. These two societies greatly helped him in his development. Later, he belonged to 'The Political Economy Club', in which he took an active part, and where he came into contact with ardent economists and others who were to guide the political thinking of the day.

In 1823, at the age of seventeen, he obtained from the East India Company, through his father's influence, an appointment in the Office of the Examiner of India Correspondence, immediately under his father. His duty was to prepare drafts of dispatches, in which he was exceptionally expert; and this continued to be his official duty until he was appointed Examiner or Chief of the Office, in 1856—two years before the abolition of the East India Company. When the moment of abolition was at hand, he was entrusted with the important task of drafting the *Petition to Parliament*—a powerful document, which Earl Grey declared to be the ablest State paper that he had ever read, standing as an example of intellectual ability and logical argumentative power for all time.

As a youth, Mill was a particularly energetic but undiscriminating propagandist of Benthamite thinking and of Radical politics. His first great literary achievements were in the *Westminster Review*, where by his pen he brought himself into a prominence that left little doubt of a brilliant literary future, and made him a real influence in philosophy and in politics. A crisis in his mental history came in 1826. It was partly owing to a breakdown in health, and took the form of a morbid mental depression, the result of unintermitted hard work and the penalty for an unwonted precocity; but it was also the emotional nature of the young man demanding fuller satisfaction than had been allowed it under the stern non-emotional training of his father. The transformation came in no slight degree through study of the poetry of Wordsworth and the philosophical lucubrations of Coleridge. The end of it all was a revolution in his nature and his mental thinking—the emergence of a new man, with a deeper sympathy, a wider intellectual outlook, a keener perception of the needs of human beings, and a realization of the importance of cultivating the emotions as well as the intellect. The change extended both to his opinions and to his character, and he himself described it as a kind of conversion. One result of this in his future philosophy was that, while still adhering to the associationist and utilitarian principles of his father and of Bentham, he was constrained to subject the Benthamite teaching to a searching examination and to amend it in various ways. His essays on Bentham and Coleridge (republished in his *Dissertations and Discussions*), glowing with a new heat and written from the heart, show how far he had advanced. This change of view gives point to his remark, in later life, on an occasion when the question was raised in conversation, of the possibility of a muster of Bentham's disciples in London at the moment, 'And I am Peter, who denied his master.'

Another potent agency in the moulding of his life, though fortunately not at the earlier and most intellectually productive part of it, was the influence over him of Mrs. Taylor, who ultimately (in 1851) became his wife, and whom he subsequently idealized as the perfect embodiment of wisdom, intellect, and character. Naturally enough, after she was gone, he writes in exaggerated terms of her qualities and virtues; but it is unquestionable that, whatever deduction may have to be made from Mill's estimate, her sway over him, from the time that the two first became acquainted, was enormous, and that she deeply affected the progress, if not the actual bent, of his thinking.

From the origination by Sir William Molesworth of *The London Review* (afterwards *The London and Westminster Review*) in 1834 to 1840, Mill occupied the position of editor (and latterly of proprietor too), and wrote many striking and important articles, giving expression to his own modification of the tenets of philosophical Radicalism and aiming at influencing for good the liberal and democratic section of the public mind.

In 1843 appeared his *System of Logic, Ratiocinative and Inductive*—an epoch-making work, which at once created an immense interest through its freshness and originality, through the power of its reasoning and the boldness of its views, through its polemic fervour and its wide knowledge, and through the even flow of its exposition which carried the reader along without jerk or jolt and cleared his mental vision at every turn. Few works of logic had been of this stamp before!

Five years later (in 1848) appeared another of his intellectual masterpieces—the *Principles of Political Economy*. Again the success was exceptional and immediate. The treatise made a profound impression, and was accepted at once by economists as of outstanding value.

Between the two came his *Essays on Some Unsettled Questions in Political Economy*. This was published in 1844, and gave a foretaste of what was to come later on. The principles of it are those of Ricardo, but the handling is Mill's own—penetrating and illuminating and marked by all the cogency of reasoning with which we have come to associate him.

His treatise on *Liberty* was begun as an essay in 1854 and was published as a book in 1859. It owed much (so he himself tells us) to his wife, and was associated sadly in his mind with her sudden and unexpected death at Avignon, in the winter of 1858–9, while they were travelling together in the South of Europe. In the same year, also, was published his *Thoughts on Parliamentary Reform*.

His *Considerations on Representative Government* was written in 1860. In 1861 appeared the papers on 'Utilitarianism' in *Fraser's Magazine*. These became the choice little treatise *Utilitarianism*, published in 1863—the fascinating work of a man as much bent on social reform as on philosophical speculation. This practical interest ought to be distinctly noted. Few works on ethics have attracted so much notice or stirred so much passion. It was extolled by some and vehemently criticized by others; but it established itself from the first as a writing which no ethical thinker could afford to ignore. By its glow and warmth, by the unconscious revelation of the writer's attractive personality, and by the sincerity of conviction with which it is written, as well as by the keenness and rapier thrust of its argumentation, it remains as inspiriting to-day as it was when it first appeared.

The *Examination of Sir William Hamilton's Philosophy* was published in 1865; and it had the merit of setting the adherents of the different philosophical schools at once by the ears. Its polemic is vigorous and keen, though not always convincing; but the book will have perennial value as one of the freshest and most suggestive expositions of experientialism or empirical philosophy to be found in the English language.

The only other writings of Mill that appeared during his lifetime to which reference need here be made are his *Inaugural Address*, on the value of Culture, in 1867, when he was Rector of St. Andrews University; and *The Subjection of Women*, which was published in 1869.

But, after his death, several treatises of engrossing interest and high import saw the light. The first was the *Autobiography*, published in 1873. The sensation that it created one well remembers. It aroused enthusiasm by the thrilling interest of the story, and by its frank and beautifully simple disclosure of the writer's own character.

A sensation not less intense was aroused by the posthumous treatise *Three Essays on Religion*, published in 1874. Mill here touched the very heart of the religious world, and, in addition, made his final contribution to the all-absorbing subject of the Philosophy of Theism. How much farther he would have gone had his life been prolonged, is only matter of conjecture; but the stage he reached, in the face of his early non-religious upbringing and without the encouragement of the fellow thinkers of his school, showed that to the end he retained an open independent mind, on which the light had never ceased to play.

The last contribution towards the fuller understanding of the man was made by the publication of his *Letters* (edited by H. S. R. Elliot) in 1910—two large volumes that bring out distinctly the wonderful multiplicity of his interest, the versatility of his mind, the singleness of his aim in the search for truth, the depth and width of his sympathies, and the diversity of his friendships.

A word remains to be said on his parliamentary career. This was brief, but distinguished. He sat as Radical member for Westminster in the Parliament of 1866–8, but failed to be returned in the succeeding Parliament. His speeches in the House of Commons were comparatively few, but his influence was great. The same mental force that characterizes his writings, the same logical power of reasoning and apt illustration, the same liberality of view and intensity of conviction, the same transparent honesty, were manifest in his Parliamentary utterances, and they produced their effect. Members listened to him attentively and respectfully, and even opponents were drawn towards him.

Gladstone once said of him in private conversation, 'When John Mill was speaking, I always felt that I was listening to a saintly man.' He was the leading Philosophical Radical in the House; but his moderation often surprised and sometimes annoyed his own party, and he held views peculiar to himself which he did not hesitate to express, even though they might give offence. Three things in special did he strenuously advocate in Parliament: the interests of the labouring classes, women suffrage, and land reform in Ireland. The last of the three he supported by his pen in his pamphlet *England and Ireland* (published in 1868), summarized by himself thus: 'The leading features of the pamphlet were, on the one hand, an argument to show the undesirableness, for Ireland as well as for England, of separation between the two countries, and on the other, a proposal for settling the land question by giving to the existing tenants a permanent tenure, at a fixed rent, to be assessed after due inquiry by the State.' In this, clearly, he pointed the way to future legislation.

II. LOGIC OF POLITICS.—In that very striking piece of work, the sixth book of his *System of Logic*, entitled 'On the Logic of the Moral Sciences', Mill has several lucid chapters on the logic of politics. By this he means the application of logical procedure to the phenomena of Society and of Government. The principles on which he proceeds are those that he had expounded in the previous parts of his treatise, especially in connexion with Induction, which involves the deductive application of laws inductively obtained. The inductive process, according to him, consists in generalizing from experience, in discovering the causes of phenomena and ascertaining their laws, according to various experimental methods specifically formulated (agreement, difference, joint method of agreement and difference, residues, and concomitant variations) and copiously exemplified, and in making application to new cases of the generalizations thus inductively reached and bringing them back to the facts of experience in order to having them verified and established. Of the inductive procedure the steps are: (*a*) Induction, dealing with facts of experience and including hypothesis, which is indispensable to scientific investigation; (*b*) Deduction, or formal inference, subsequent to and reposing on induction; and (*c*) Testing by experience or verification. In dealing with social phenomena, the utmost stress is laid on the third point in the process—the need of verification, without which we should be condemned to mere guess or conjecture.

In the realm of social science, at that time dominated by political thinking, Mill's object is to show where the ordinary politician is apt to fail in his reasonings and why; and to bring out and enforce what he regards as the only effective mode of coping with phenomena so complex and peculiar as those that lie before him. He has first of all to meet two kinds of political reasoners who err, though in different ways, from applying an inapplicable and therefore an ineffectual method; one (the less-instructed politician) trusting to specific experience, and the other relying on abstract thinking. The mode of the first is the undiscerning application of the various experimental methods to society, oblivious of the fact that it is not in our power to experiment to our hearts' content with society, so as to vary the circumstances in the way required for scientific elimination; nor can we exhaust the causes of complex social phenomena (say, the prosperity of a country), or reduce the causes to one. We are really dealing with 'composition of causes'. In the case, for instance, of Protection and its influence on national wealth, it is dangerous to argue unchecked, according to the experimental methods alone, from the flourishing condition of a foreign protected nation to the wisdom of restrictive legislation for Great Britain, whose circumstances are in many ways entirely different. This first mode of insufficient reasoning he designates, in his own peculiar terminology, 'The Chemical, or Experimental Method'.

The method employed by the other type of politician (and no less inadequate, though appealing to certain high political thinkers) is what he calls 'The Geometrical, or Abstract Method'. It rightly starts with the laws of human nature, human nature as we know it in experience, and recognizes the necessity of a deductive application of them to the phenomena in hand, but errs in accepting geometry as the type of all deductive reasoning. Obviously, in society, where we have to take progress into account, where the phenomena are ever changing and conflicting, and where it is necessary to consider *tendencies*, geometry is not elastic enough to cope with the situation. His great example is the 'Interest-philosophy' of the Bentham school, which he dispassionately criticizes, notwithstanding that he was himself an adherent of the school. His father's presentation of it is what is naturally uppermost in his mind. He canvasses each of its two central positions—that 'the actions of average rulers are determined solely by self-interest' and that 'the sense of identity of interest (of the governor) with the governed is produced and producible by no other cause than responsibility'. He does not regard either of these propositions to be true, and the second of them he maintains to be extremely wide of the truth.

He is thus thrown upon a third method, which alone is regarded by him as sufficient for sociology. His distinctive name for it is 'The Physical, or Concrete Deductive Method'. It 'proceeds (conformably to the practice of the more complex physical sciences) deductively indeed, but by deduction from many, not from one or a very few, original premises; considering each effect as (what it really is) an aggregate result of many causes, operating sometimes through the same, sometimes through different agencies, or laws of human nature'.

Sociology is thus, in Mill's view, a system of deductions from the primary laws of human nature. But if so, it cannot be, in the strict sense of the phrase, a science of positive predictions. Yet science requires (so Comte had laid down), as one characteristic, that it shall have the power of prevision. The difficulty is surmounted by the fact that Sociology, though unable to make absolutely certain predictions, in the way that astronomy, for instance, does, can nevertheless *gauge tendencies*; and knowledge of tendencies, insufficient for strict scientific prediction, is of the utmost value for *guidance*—which is enough for the legislator and the politician.

Allied to this is another point. A peculiarity that attaches to Sociology is the fact that man and society with which it deals are progressive. As age succeeds age, experience grows, and one generation of men differs from another. At any particular time, the circumstances of a society determine its character; but the circumstances are in turn reacted on by the society and in no small degree are moulded by it. This throws the student of social science upon the study of history, so that he may discover, if possible, the law of human progress. But the generalizations made from the facts and events of history are not in themselves sufficient to guide us; they are simply empirical laws of society, and need to be verified. This verification can be effected only by connecting them with the laws of human nature, as revealed in experience, and by showing deductively that such derivative laws are precisely those that we should naturally expect as the consequence of those that are ultimate. This method is what Mill calls (somewhat enigmatically) 'The Inverse Deductive, or Historical Method'.

In his handling of Sociology, Mill shows how greatly he was influenced, at the time of his writing, by the French thinker, Auguste Comte. To Comte is owing the recognition of Sociology as a distinct science, and the clear enunciation of the scope and method of it. The Comtian position is reproduced by Mill, and much is made of Comte's distinction between 'Social Statics' and 'Social Dynamics', as constituting the two branches of social science; the first having for its province social *order*—the observation and examination of different states of society, and the discovery thereby of the requisites and grounds of social union and stability; and the second considering social states as succeeding each other in time, and aiming at ascertaining the laws of social *progress.* Later, in his *Representative Government*, Mill is explicit on the point that 'order' and 'progress' cannot really be separated, much less opposed, but that the distinction between them is mainly one of simplification of the problem and expository convenience. As to social progress, it is evident that it depends on the varying range of men's knowledge and the nature of their beliefs. Every great change of a social kind is preceded by a change in people's conceptions and convictions; so that social progress is amenable to invariable laws. This is the fact that lies at the root of a philosophy of history, which is 'at once the verification, and the initial form, of the Philosophy of the Progress of Society'. Help is also given by a study of Statistics—a subject that had, just at the moment of Mill's writing, been brought into special prominence by the startling speculations of Buckle.

Mill's presentation of social science threw new light upon the subject to English thinkers, and created a widespread interest in America as well as in Great Britain. Much has been done in Sociology since his day, but that does not detract from the value of his contributions. If he did not make sufficient use of the conception of social *evolution* (as we find it in Benjamin Kidd, for instance), it can only be said that, perhaps, the conception of social evolution has too great a strain put upon it at the present time.

III. ETHICS.—The current ethical philosophy of Mill's day based moral ideas and principles on intuition, and appealed to a distinct moral faculty or conscience as the ground of moral decisions, instead of making an appeal to experience. Regarding them as native to the human constitution, it clothed them with a superior dignity or worth, and conceived them as beyond the range of criticism: they had simply to be accepted by us, unquestioned and unanalysed. It strenuously opposed such a test as that of utility or pleasure, and practically claimed for itself a monopoly of the notions duty, virtue, obligation, right. Against this Mill urged the doctrine of utilitarianism, and set himself to prove that the utilitarian is as strongly on the side of conscience, duty, rectitude, self-devotion, as any intuitionist; and that 'the terms, and all the feelings connected with them, are as much a part of the ethics of utility as that of intuition'. He, further, made an appeal to his opponents to discard prejudice and try to do justice to the utilitarian conception of happiness; which stood, not for the individual agent's happiness alone, but for that of all concerned, and which embodied the enthusiasm of humanity—a noble sentiment, specially associated with the Christian religion and embodied in Comtism. 'As between his own happiness and that of others, utilitarianism requires him to be as strictly impartial as a disinterested and benevolent spectator. In the golden rule of Jesus of Nazareth we read the complete spirit of the ethics of utility: To do as one would be done by, and to love one's neighbour as oneself, constitute the ideal perfection of utilitarian morality.' 'I regard utility', he says in the treatise *On Liberty*, 'as the ultimate appeal on all ethical questions; but it must be utility in the largest sense, grounded on the permanent interests of man as a progressive being.' Accordingly, he did not deny the existence of moral intuitions, but he raised the question of their value; and, in working out an answer, demonstrated that many notions and principles in morals that passed for intuitive were falsely so regarded, being merely common opinion and belief, or, it might be, mere sentiment or prejudice, untested by reason. On the other hand, he maintained that what is valuable in morality is founded in experience and may be corroborated by it. Thus only is morality satisfactorily established, and a science of ethics rendered possible. The moral feelings are not innate but acquired; yet they are not, on that account, the less natural to man.

Instead, then, of founding moral ideas on intuition and removing them beyond the reach of experience, he brought them into direct relation with experience and insisted on testing their validity thereby. In his completed analysis, he found, as Bentham and James Mill had done, that they ultimately derived their character from the pleasure-value of their consequences.

Nevertheless, although pleasure is the ultimate test of moral value, Mill thought it necessary to strengthen the utilitarian position by drawing a distinction between the different kinds of pleasure. This is a distinct departure from the position of Bentham and James Mill. To them, pleasures differ only in quantity, and one pleasure is as good in itself as another: as Bentham put it, 'Quantity of pleasure being equal, push-pin is as good as poetry.' But now J. S. Mill regarded it as quite compatible with the principle of utility to introduce into the conception the distinction between quantity and quality of pleasure, and to lay emphasis on quality. Pleasures are thus conceived as in themselves intrinsically different: there are higher and lower among them.

E106

The proof of this is appeal to intelligent people who have had experience of both : *they* are the judges, and their testimony is decisive. 'It is better to be a human being dissatisfied than a pig satisfied; better to be Socrates dissatisfied than a fool satisfied. And if the fool, or the pig, is of a different opinion, it is because they only know their own side of the question. The other party to the comparison knows both sides.'

No doubt, this recognition of a qualitative distinction among pleasures strengthens the theory of morality; but that it is logically legitimate, on strict utilitarian principles, is not quite obvious.

Mill's insistence on pleasure-value as determining morality comes out strikingly in his treatment of disinterestedness and of virtue. No intuitionist could write of disinterestedness and virtue in more glowing terms, nor could any one express himself with intenser conviction. Yet, to him pleasure lies at the root of them—disguised, but operative.

The explanation of this seeming inconsistency is to be found in Mill's associationist psychology. The transformation of egoism (or regard for self) into altruism (or regard for others) is effected, he thinks, by the same process that turns the money-seeker into the miser. A man may begin by desiring money for the pleasure that it can procure him, but by and by he transfers his affections from the end to the means, and goes on amassing money for its own sake. So, although pleasure to self is the ultimate explanation of human disinterestedness, self is forgotten by the ethical man in his earnest and devoted service of others : his highest pleasure comes when he does not directly seek it; his own happiness is found in doing good to his fellows or to other sentient creatures. Mill lays great stress on what, since his day, has come to be known as the paradox of pleasure—the Hedonistic Paradox. Directly aiming at pleasure may fail to secure it: as Bain puts it, 'Happiness is not gained by a point-blank aim; we must take a boomerang flight in some other line, and come back upon the target by an oblique or reflected movement.' This paradox fits in with Mill's doctrine, but, unless carefully guarded, may mislead us; for pleasure is frequently got (as at the dinner-table) by directly aiming at it, and nothing would be gained by concealing from ourselves our object in pursuing it.

Virtue, in like manner, is regarded by Mill as explained by association of means and end. It is a means to happiness; but it is also desirable in itself and so is an 'ingredient' in happiness or a 'part' of it. Yet, in the ultimate analysis, it is found that 'those who desire virtue for its own sake, desire it either because the consciousness of it is a pleasure, or because the consciousness of being without it is a pain, or for both reasons united'.

It was a noteworthy feature of Mill's teaching that he saw, in a way that neither Bentham nor James Mill had done, the essentially social nature of morality; and, even more than this, the stimulating and uplifting fact that society itself has a moral end—the moral good of its members. Justice and sympathy are the bulwarks of it. In other words, the most potent factor in right conduct is the social feelings, or the desire to be in unity with our fellow men; and our social inclination is not stationary and stereotyped, but may be cultivated and developed and becomes stronger, without being expressly tended, from the influences of advancing civilization.

From the position that pleasure or happiness is the sole object of desire, or that every individual desires his own happiness or pleasure, Mill passes to the farther position that the individual should desire and promote the general happiness. 'No reason can be given why the general happiness is desirable, except that each person, so far as he believes it to be attainable, desires his own happiness. This, however, being a fact, we have not only all the proof which the case admits of, but all which it is possible to require, that happiness is a good: that each person's happiness is a good to that person, and the general happiness, therefore, a good to the aggregate of all persons.' The reasoning here is said to be invalid; for, it is maintained, you cannot argue from the fact that a person actually desires a thing to the desirability of that thing; nor can you pass, except by a stealthy transition, from the individual's own happiness to that of his fellows. This last is known in Logic as 'the fallacy of composition'. But Mill did not regard the individual as a purely isolated unit; he conceived him as essentially a member of society, with strong social instincts, sympathies, and feelings; so that, in *his* desire, we have so far represented the desire of his fellows, and any desire (like that of happiness or pleasure) that is shared in by mankind in general may very well be regarded as a *natural* desire, and, therefore, be trusted. The objector forgets to take into account Mill's reiterated emphasis on the need for regarding the individual as sympathetic with others, as co-operating by nature with them, and as living under a social order—which means mutual interest and goodwill. Here he is on very strong ground; but his position would have been stronger had he referred

to Heredity as a social factor. In answer to a correspondent, in June 1868, he explained his position thus: 'When
I said that the general happiness is a good to the aggregate of all persons, I did not mean that every human
being's happiness is a good to every other human being,
though I think in a good state of society and education
it would be so. I merely meant in this particular sentence
to argue that since A's happiness is a good, B's a good,
C's a good, etc., the sum of all these goods must be a
good' (*Letters*, II, p. 116).

Mill's psychology, and more still his logic, led him to
take a deterministic, but not a fatalistic, view of man's
will. That every individual man is a part of nature and
is subject to the laws and uniformity of nature, was to
him incontrovertible. But that does not take us far.
The individual is also a person, and counts for a living
and efficient force in nature, and social and moral
development would be impossible were it otherwise.
Yet, this does not mean that man's will is 'free' in the
popular acceptation of that term—that he is at liberty
to choose, if not to act, precisely as he pleases. Nor does
it mean that the will is necessitated or coerced, in the
sense that the extreme necessitarian attaches to the misleading term 'necessity'. It means that he is subject to
the law of causation, just as the physical world is. But
what is 'causation', and what is 'cause'? In Mill's
acceptation, as laid down in the *Logic*, causation does not
involve *must*—there is no bond or nexus or coercive
necessity between cause and effect: it expresses simply
uniform and unconditional sequence. He looks upon a
cause as the sum total of conditions issuing in the effect.
It is that which uniformly and unconditionally precedes

an effect: or, as Bain puts it, 'which happening, it happens; and which failing, it fails'. In this way, man's choice may be said to be determined, in the sense that there are conditions indispensable to its occurring. A man has a character; and his character is not inexplicable, fluid, capricious, but stable or comparatively fixed and formed—something on which we may calculate or depend. He himself is a rational being, acted on by motives, impelled by desires; and a motive is a cause, inasmuch as it is indispensable to his choice. At the same time, a person's desires in large measure determine his character, and thus his character is his *own*—made *by* him, not *for* him. 'Given the motives which are present to an individual's mind, and given likewise the character and disposition of the individual, the manner in which he will act might be unerringly inferred—if we knew the person thoroughly, and knew all the inducements which are acting upon him, we could foretell his conduct with as much certainty as we can predict any physical event.' 'If necessity means more than this abstract possibility of being foreseen; if it means any mysterious compulsion, apart from simple invariability of sequence, I deny it as strenuously as any one in the case of human volitions, but I deny it just as much of all other phenomena.' That is all that 'determinism' in ethics means to Mill.

These ethical positions lie at the basis of Mill's political and economic philosophy; and we shall find them cropping up frequently as we proceed.

CHAPTER IX

J. S. MILL

POLITICAL ECONOMY; PSYCHOLOGY AND THEORY OF
KNOWLEDGE; WOMEN'S RIGHTS

I. POLITICAL ECONOMY.—Mill's economic speculations
are of a piece with his ethical teaching. In either case,
he is dealing with a science; and a science is possible
only if there is a power of forecasting actions or predicting
events. We must be able to gauge tendencies, even when
we cannot be absolutely certain of results; and if we
had no prevision of consequences, we should be condemned to bare conjecture or surmise. That is what Mill
meant by taking a deterministic view of human will in
his utilitarian ethics. Determinism as there applied
means only that we can foresee volitions, or judge
beforehand how character will reveal itself. The same
is applicable to Political Economy; the possibility of
which as a science simply lies here, that we know that
men are moved by a desire of wealth, and that this
actuating principle, if we suppose it to be dominant, will
manifest itself in such and such ways and lead to such
and such consequences. Although it is not the direct
purpose of political economy to inculcate ethics (its
subject-matter is wealth and the laws and modes of its
production, distribution, and consumption), it is impossible to treat of wealth in entire abstraction from the
nature, motives, and social character of the human

beings implicated—beings that are amenable to reason and whose lives are formed on a plan. The individual works for self, but he works for others also and for the good of the community. Inculcation of these facts goes far to redeem economic science from the charge (not groundlessly made against the earlier economists) of being 'the dismal science'. Dismal it behoved to be so long as its exponents laboriously elaborated the principle —'Sell in the dearest market, buy in the cheapest'; but, when due cognizance was taken of the circumstance that, after all, the economist is not concerned with stocks and stones but with beings of flesh and blood, possessing feelings as well as thoughts, and bound to each other by ties of family and race, a fresh vitality was infused into the subject.

It was peculiar to Mill, as compared with his British predecessors, that he widened the conception of Political Economy. Influenced by Comte, though not by him alone, he viewed it as inseparably associated with the philosophy of society, thereby conjoining consideration of economic theory and principles with that of their social applications. It was this combination of theory and application in Mill's treatise on Political Economy that gave it its immediate and wide popularity. People felt that such a science might well claim to be a practical guide to them when it did not ignore, but explicitly recognized, the other branches of social philosophy, and took into account their interconnexion and interdependence.

In his role of democrat, Mill reviewed the current doctrine of private property and inheritance and the standing problem of the ownership of the land. As to property, he held that the individual is entitled to the use of his own faculties and to whatever he can produce thereby, and that he has the right to bequeath what is his own to another, who, in turn, has the right to accept and to enjoy it. Property is a social institution necessary to the good and progress of mankind, and not something thrust by fraud upon the majority with a view to keep them in servile obedience and comparative poverty in the interests of the rich and powerful and unscrupulous minority. Hence, contrary to the teaching of Socialism, inequality is a social necessity. But these rights are subject to many limitations—such as, the existence of children whom a father has to provide for, whose claims to part of the paternal possession may override those of any one outside the family circle to whom the father may have bequeathed his fortune. In like manner, private property in land, with due limitations, is justifiable. For land is valuable only when it is cultivated and made productive, and that means expenditure of capital upon it; and as the outlay, as a rule, is not immediately remunerative, but brings return only after a time (it may be years of waiting), there would be no inducement to the capitalist to make improvements and incur outlay unless he had a sufficient period secured to him in which to reap the benefit—the most potent inducement being perpetual tenure.

In these respects, i.e. with regard to private property in general and to ownership of land in particular—Mill accepted the social arrangements and institutions of the country at the time, and merely set himself to mitigate the inequalities consequent upon them. He advocated such things as the abolition of primogeniture, modifying the system of entails and restricting the right of bequests, and, in the case of Ireland, whose condition then as now was perplexing to the legislator, the creation of peasant proprietorships. He also safeguarded his teaching on property in land, and was careful to inculcate that landed proprietors are public functionaries, and so that the State is at liberty to deal with the land as the general interests of the community may require, even to the extent, if need be, of doing with the whole what is done with a part whenever a Bill is passed for a railroad or a new street. His last pronouncement on the land question was made the very year that he died (1873), and was the last thing that he wrote. It was a brief exposition of the conditions (with answer to objectors) under which proprietorship in land is legitimate, and a defence of the right and the duty of the State to look after and lay hold of the unearned increment.

As time went on, and Mill's sympathy with the labouring classes was more and more brought out, as he came more vividly to realize the hardship and, as he conceived it, the injustice, that a few only should be born to riches and the vast majority to poverty, he became increasingly drastic in his proposals for reform. The evolution of his opinions was this: Earlier in life, he aimed chiefly at enlightening the opinions and changing the habits of the manual labourers. For this purpose, he laid great stress on education, and preached the necessity of ceasing to keep the labouring classes in a state of patriarchal dependence on the rich, and of teaching them to think and to determine for themselves. Their destiny was in their own hands; and what they wanted was justice and self-government. Given education and increase in intelligence, together with the love of independence, and there would inevitably come, he thought, improvement in habits and mode of living, and an effective appreciation of the need to keep offspring in due proportion to capital and employment or the means of support. Prudence would rule, and Malthusianism would be seen to be the true wisdom. Much would be effected also by the unrestricted opening up of industrial occupations to both sexes.

All this went a considerable way, but Mill proceeded farther. He had at first opposed Socialism, or, at the best, treated it critically and coldly; but now, after much study and meditation, he saw virtues in it which had previously been concealed from him, and, with the openness of mind that characterized him, he readily welcomed them. With extreme socialism he had not sympathy, even to the end: he never, for example, advocated the nationalization of the land. In general, he did not accept the socialism that swamped the individual; nor did he share the socialist's dislike of competition: he regarded competition as essential to successful trade and as a security against the evils of monopoly. But he approved of Trade Unionism, and the socialist's idea of voluntary co-operation he accepted unreservedly. In the famous chapter in the *Political Economy* (from the third edition onwards) that portrays the future of the labouring classes, he strongly insists on the unsatisfactory character of the opposition between capitalist and labourer, and maintains that 'capitalists are as much interested as labourers, in placing of operations of industry on such a footing that those who labour for them may feel the same interest in the work which is felt by those who labour on their own account'. Hence, he holds the system of large industrial enterprises to be the best for labour itself, not only because it gives a larger return for the labour employed, but because it means co-operation and association of individuals, people bound together by a common interest, thereby encouraging public spirit, a sense of justice, equality, and generous sentiments. He held the ideal and probably the ultimate goal of the working man to be partnership in one of two forms—association of labourers with the capitalist, or association of labourers among themselves.

While thus distinctly socialistic, however, Mill still clung to his individualism. He was very jealous of the interference of the Government in economic and industrial matters. That there must be State control to a certain extent is inevitable and right. But State interference in the business of the community should be restricted to the narrowest compass: individual free agency, personal liberty and spontaneity, must be protected with the utmost possible rigour. Hence, the principle that he lays down regarding the limits of the province of Government is: '*Laisser faire*, in short, should be the general practice: every departure from it, unless required by some great good, is a certain evil.' In other words, his maxim was: Let people look after their own business—as being immediately interested in it, they are likely to attend to it best; let the Government intervene only in the interests of the community in general, and as seldom as possible.

These are characteristic points in Mill's handling of economic and social questions. How he deals in detail with the various sections of political economy, as definitely formulated for the student of economics, in his great treatise, is too wide a subject for our space. Suffice it to say that he traverses the whole range of the science (labour, capital, wages, rents, profits, rate of interest, international trade, value, money, credit, taxation, &c.), and raises questions and initiates discussions that are both fruitful and engrossing. His logic aids him in the admirable arrangement of topics, and his freshness of mind and clearness of style give distinctive character to his handling: his insight and his wisdom seldom fail him. His work stands as a landmark in economics.

II. PSYCHOLOGY AND THEORY OF KNOWLEDGE.—Although pre-eminently a psychologist, Mill has no formal treatise on psychology. His teaching is contained in sections of his *Logic* (such as those that deal critically with the views of Whewell and Herbert Spencer regarding the Inconceivability of the Opposite as the criterion of Truth), is partly developed in the *Utilitarianism*, and finds its fullest expression in the *Examination of Sir William Hamilton's Philosophy*, and in Mill's Notes in his edition of his father's *Analysis of the Phenomena of the Human Mind* (1869). Under Ethics, we have already seen his treatment of desire, and his doctrine of human will. To these we need not return. It remains only to indicate the leading points in his treatment of intellect and of the problems connected with it.

The intellectual controversies of the schools are mainly connected with the origin or genesis of knowledge (the place and value of intuition and experience respectively); the nature and meaning of external reality, or the doctrine of sense-perception; the nature of the mind, self or ego—of the subject as contrasted with the object, of mind as opposed to matter; and the meaning of the Absolute.

The question regarding the source and validity of knowledge resolves itself into this: Is the whole of man's knowledge explicable by experience, or is there knowledge (or elements of knowledge) that experience is incapable of accounting for, that is presupposed in experience, and is, therefore, *a priori* or native to the mind itself? Mill and the school to which he belongs answered unhesitatingly that experience, working by association, is fully competent to explain knowledge in all its kinds or forms, and that intuition, though a fact, derives the power that it possesses from experiences indefinitely repeated and uncontradicted. Their opponents, on the other hand, maintained that, although experience has much to do with knowledge—is, indeed, indispensable as the condition of its expression and development—there are knowledges that show characteristics which experience cannot explain, as it could not have originated them. These characteristics are *necessity* and absolute *certainty*, which no amount of experience can produce, but which originate in the mind itself. Experience can give the *is* and attest the *probable*, but cannot generate the *must* or the *ought to be*. How, for instance, explain by experience alone the axioms of mathematics, or the principle of causality? It cannot be done—so said the intuitionist.

It was Mill's function, for which he regarded himself as specially fitted, to demonstrate in what way these necessary truths, with the consequent certainty of belief, were generated by experience, according to the laws and working of association; and thereby to vindicate the value of the experiential philosophy which he had inherited from his father, and get rid of the intuitionism that held sway in British philosophy, which he regarded not only as illegitimate and illusory, but also as the bulwark of irrational prejudices (social, ethical, religious, political) and a hindrance to intellectual liberty and progress. 'The notion that truths external to the mind', he said, 'may be known by intuition or consciousness, independently of observation and experience, is, I am persuaded, in these times, the great intellectual support of false doctrines and bad institutions. By the aid of this theory, every inveterate belief and every intense feeling, of which the origin is not remembered, is enabled to dispense with the obligation of justifying itself by reason, and is erected into its own self-sufficient voucher and justification. There never was such an instrument devised for consecrating all deep-seated prejudices.' It will be noted that social institutions, as well as intellectual beliefs, are regarded as coming under the sanction of intuition. This gives additional point to Mill's polemic.

Intuitions, then, must be tested: that is Mill's position. But if they are subject to being tested, they cannot be ultimate or in themselves indisputable. The test must necessarily be experience. Those of them that stand this test are to be accepted; those that fail under the application of it, we reject.

But while thus combating intuitionism, as understood and applied at the time, Mill had to face the problem of mind and matter—the nature of the external world as perceived by us. Here, too, his position is distinctive. External reality is given us in and through sensations; and apart from sensations (actual and possible) it has for us no meaning. An external object is a group or congeries of attributes, which we obtain from the various senses, and associated in a peculiar way; and besides this, there is nothing—there is no 'substance' or underlying thing, or 'thing-in-itself', as the metaphysician maintains. Matter *by itself* is a meaningless phrase. What we have in the perceptive process is actual sensations, sensations of something here and now present, with the *expectation* of *possible* sensations, when the actual are in abeyance. For example, in presence of this chair, I immediately experience certain sensations (of sight, touch, muscularity, &c.), which constitute the chair to me. When I withdraw from the room, I believe that, if I came back, I should experience these same sensations. I do not regard the chair as going out of the room along with me, but imagine the sensations I now have as still possible, and realizable by me on my return. This combination of actual and possible sensations, coupled with expectation, is precisely what I understand by the *permanence* that characterizes external objects, as distinguished from inward feelings, which are fleeting; and it explains also what I mean by their *externality* or independence of me. Accordingly, Mill defines matter as the *permanent possibility of sensation*. It is no mysterious and unknowable something, but simply my idea of certain sense-experiences under given circumstances and in definite modes. To suppose a chair existing apart from all its perceptive qualities (hardness, colour, form, &c.), is to suppose it still existing *as a chair*. But to exist as a chair, it must possess the qualities of hardness, colour, form, &c. And so the supposition of an existent quality-less chair is absurd.

This doctrine of the external world is known in philosophy as *Psychological Idealism*.

But, besides matter, there is mind. It, too, according to Mill, is dependent for its being on experience and association, and is conceived by him as the *permanent possibility of feeling*. He allows, however, that there is in mind or self or ego something more than what there is in matter. It is characteristic of mind that it is not only a series of states, but is also *conscious* of them. In other words, the ego or self is that which binds the states together, just as the thread holds the beads together in a necklace. *How* this can be, is utterly inexplicable; but we must accept it as we find it.

True to his experiential standpoint, Mill lays stress on the relativity of human knowledge—on the necessity of regarding the object known in relation to the subject knowing: object implies a subject, and only in so far as the subject has faculties adequate to cognizing the object is the object known, although the object may have other qualities that would be apprehended by a different or a higher subject, or by the same subject endowed with a greater number of faculties. Thus Mill obtains his doctrine of the Absolute or God. As knowledge is essentially relative, there is no such thing as 'the Absolute', if by that be signified a self-existent unrelated something, to which no attributes or properties can be ascribed, in the sense of attributes and properties as conceived by us. Any Absolute that can be regarded as having for us a meaning is that which may be, in part at least, apprehended by human intelligence and characterized in terms of human experience, however much its (or His) attributes may surpass those of man. Sir William Hamilton's teaching, and more especially the reproduction of it by his disciple Dean Mansel, is what Mill has here specifically in view. An inconceivable and unknowable Being, such as Dean Mansel, in his Bampton Lectures on *The Limits of Religious Thought*, had represented God to be, is a contradictory conception. The idea of God as a Being that may not be spoken of as knowing or loving or willing, or conceived as possessing moral attributes, or be regarded as having anything in common with man, and yet must be accepted by us as an object of faith and worshipped without question, is worse than nonsensical—it is mischievous and 'profoundly immoral', inasmuch as it makes slaves of us and breeds hypocrites. The only safe basis of the higher speculation is human experience. The Absolute must be Absolute *to us*, and be characterized by qualities (unless the term is to be a mere sound) that are significant of qualities which we ourselves possess, though in a lower or imperfect degree (wisdom, goodness, justice, love, &c.), and such, therefore, as we can in part understand.

Thus the questions of the higher speculation are real living questions to Mill (and not mere subjects for clever academic debate), with practical bearing. He finds a real meaning in the dictum, 'Man is the measure', and may be taken as the precursor of William James and his fellow Pragmatists.

III. WOMEN'S RIGHTS.—Mill was early impressed with the social and legal disabilities in life under which women suffered, as compared with the privileges and freedom of men. Law and legislation, supported by public opinion, had all along proceeded on the assumption that practically only one sphere of usefulness was open to a woman— that of married life, with the management and charge of a household, including the bearing and bringing up of children. But even here, and indeed here most glaringly of all, the law had emphasized the inferiority of women by so extending the powers of the husband and safeguarding his rights as to make the married woman's lot, to all intents and purposes, one of slavery. Outside the sphere of married life, there was comparatively little left for a woman to do. If she preferred to remain single, her usefulness was miserably circumscribed. She was not allowed to be educated to the extent or in the way that a man was—she was shut out from the Universities and the higher learning: she was prevented from taking any active part in public life, and was discouraged from even interesting herself in public affairs; most of the high posts and prizes were kept jealously beyond her reach; and the idea of her being permitted to exercise the vote at a political election, much more of her aspiring to guide public sentiment in things political, or of herself ultimately occupying a seat in Parliament, seemed so outrageous as to be almost unthinkable. That was the mid-Victorian state of affairs, and people acquiesced.

Mill's sense of justice was stirred on the contemplation of this, and he set to work to advocate the removal of these inequalities, which he conceived to be gross injustices, and to aid the cause in every practical way that he could. He was eager to 'emancipate' women (so he phrased it) and to put an end to their 'subjection'. He claims to have been the first to plead their cause in Parliament, and to urge their enfranchisement; and he fanned an organized public agitation in their favour. He was very intimately connected with the London Committee of the Society for Women's Suffrage, and at a critical moment in 1871, he was the means of saving it from disaster. This service he rendered through the active instrumentality of Professor Croom Robertson, of University College, London—as is amply shown by letters of Mill to Robertson in the possession of the present writer. The peril lay in an attempt, by extreme members of the Committee, to associate the cause of women's franchise with that of the repeal of the C. D. Acts.

Mill's unqualified advocacy of women's rights was the natural outcome of two principles of human nature that he held strongly—(a) the indefinite modifiability of character (see *Autobiography*, p. 108), and (b) the power of outward circumstances to determine differences among human beings (*Logic*, Book VI, Chapter V, §3). If differences between the sexes (apart from certain obvious physical and physiological differences) are due solely to external circumstances, then they are removable. Consequently, he deliberately set himself to prove that these differences are not fundamental and inevitable, but had been created by man's long-continued usage of women in one definite line, and that, if a different state of matters were brought about and women obtained social and political freedom, they would disappear. As they had arisen from the law of force having been uniformly employed, instead of the law of justice, they would cease whenever the application of force ceased and justice became supreme.

The arguments on which he rested his case were such as these: The position of women in the land is altogether anomalous. Theirs is almost the solitary instance of persons being excluded from high offices and functions in the community solely by the fatality of birth. The case of Royalty in Great Britain is hardly an exception. For, though the office of King is hereditary, the law of the land so circumscribes the sovereign's functions as to throw the real power into the hands of his chief minister, who does not inherit his position, but obtains it by his own merits. Again, the exclusion is itself unjust, because it has not the sanction of experience to recommend it. Not only have women, in large areas of thought and of action, never been allowed to test their ability at all, and so to show the quality of it, but the significant fact that in spheres where they have had opportunity to develop freely they have proved themselves at least the equals of men, has been ignored or its significance minimized. Such treatment is contrary to the progressive and enlightened movement of modern times. The great principle that actuates modern social thought is that birth is no barrier to a person's advance (material or intellectual), that no one is really *born* to a particular trade or calling or profession, out of which he cannot raise himself if he be dissatisfied or if he have aspirations, but that each is born free and is at full liberty to choose and make his position in the world for himself with what advice from others he cares to take, and according to his ability to seize the favouring opportunity. 'A fair field and no favour' (to use a homely phrase) is the condition of progress, accepted and allowed everywhere, except in the case of women, and the exception is unjust and tyrannical. Again, opposition to the removal of women's disabilities, social and political, arises mainly from custom and prejudice. When it is said by the opponent that it is *unnatural* for women to throw themselves into political movements and to wish to rival men in the various occupations and professions that have hitherto been restricted to the stronger sex, the meaning really is that it is not *customary* to do so—that it is a breach of what is usual and established. But custom is not sacrosanct in the eyes of reason, though it acts as a powerful force against reason. 'What is just?' and 'What is expedient?' are the real questions at issue; and the decision should be arrived at on the merits of the case. 'Neither does it avail to say that the *nature* of the two sexes adapts them to their present functions and position, and renders these appropriate to them. Standing on the ground of common sense and the constitution of the human mind, I deny that anyone knows, or can know, the nature of the two sexes, as long as they have only been seen in their present relation to one another. . . . What is now called the nature of women is an eminently artificial thing—the result of forced repression in some directions, unnatural stimulation in others. . . . In the case of women, a hot-house and stove cultivation has always been carried on of some of the capabilities of their nature, for the benefit and pleasure of their masters.'

By these and suchlike arguments Mill pressed home the necessity of a full and speedy emancipation of women. And he strongly urged that the beneficial results of the reform would be numerous and far-reaching. Not only would the liberated women themselves be happier (a sense of freedom alone gives happiness), not only would an atmosphere be created in which their natures could more easily and spontaneously develop and their activities expand, but the community also would greatly benefit. In the reign of justice, when employments and posts were thrown open to unrestricted competition, the mass of mental faculties available for the higher service of humanity would be doubled, and the level and efficiency of the service presumably raised; and there would be an increase also in the humanizing influence of women on the general mass of belief and sentiment. And, further, society would benefit by the *practical* turn of the mental capacities characteristic of women—their intuitive perception, their quickness of apprehension, and their nervously sensitive nature.

There can be no question of the cogency of much of Mill's reasoning; and it produced a marked effect. Indeed, it would hardly be too much to say that the higher education of women in Great Britain at the present moment, and the increased opportunities now afforded on all hands for the exercise of their practical talent, in social and other spheres, are largely owing to his advocacy and lead.

But, though Mill's reasoning is in great measure cogent, it is not completely so. It is the reasoning of the special pleader, who puts too great an emphasis on some points and passes lightly over others. For one thing, his sombre picture of married life is overdrawn, and his conception of the home inadequate. The word 'subjection' as applied to the state of women here, although appropriate to a certain number of cases, and to certain legal aspects of the marriage contract, is altogether too strong, if unqualified, in face of the high ideal which marriage implies, involving the unreserved mutual love and confidence of the parties immediately concerned. The 'commanding' and the 'obeying' in the case (the first by the husband, and the second by the wife) is not founded on the spectre of 'force' lying at the back of it, but on the union of hearts, which belongs to an entirely different sphere. Nor has Mill a sufficiently high appreciation of the place and functions of the mother in a home, and the vast nobility and importance of her work in the upbringing of her family, not only for the members of the family itself, but for the first and highest interests of the State and of the world.

Again, he does not sufficiently realize the magnitude and the significance of the physiological differences that are involved in the distinction of sex—differences of nature, which cannot be obliterated by legislation (he regards sex merely as an 'accident'); nor does he realize the evils that would attend a radical alteration in the relations between men and women. What of chivalry, if women were plunged absolutely into 'the turmoil of masculine life'? And if chivalry goes, the nature both of men and of women would suffer. What of the special attractive feminine traits and graces—which, if lost, would be nothing less than a calamity to the world? Mr. Gladstone put it well when, in 1892, he said, in his opposition to the proposal to extend the parliamentary suffrage to women: 'I have no fear lest the woman should encroach upon the power of the man. The fear I have is, lest we should invite her unwittingly to trespass upon the delicacy, the purity, the refinement, the elevation of her own nature, which are the present sources of its power.'

Lastly, Mill, in his one-sided view, is too apt to confuse 'inequality' with 'injustice', and to think that 'subordination' must needs in itself be an evil. But subordination is necessary in life, if society is to exist and go on at all; and only that kind of subordination is reprehensible that is not founded on worthy superiority.

CHAPTER X

J. S. MILL

I. LIBERTY, OR A PLEA FOR INDIVIDUALITY.—Deeply impressed with the fact that social and political progress depends largely on the originality and energy of the individual, and not less concerned with the tendency in the democracy to swamp the individual in the general, Mill stood forth as the advocate-in-chief of individuality —of the supreme importance of developing the individual in all the completeness of his being, so that his active and his intellectual nature might have their utmost scope and reach their highest efficiency. Without this, he thought, general progress was impossible. He fully recognized that there is a kind of individuality that ought to be suppressed, viz. unbounded and unrestricted liberty to develop oneself regardless of one's social duties and responsibilities. But, this apart, democracy itself (which was imposing the tyranny of collectivism) and the public good demanded that every encouragement should be afforded to each person to assert himself in his own peculiar way, so that his services might be as great as possible and the world enriched by 'variety' of character. That, Mill believed, was precisely what social opinion and State legislation under democracy were refusing to do. They were not only demanding co-operation of the individual with his fellows, but were doing their utmost to reduce him to a common type, to absorb him in 'sociality', and were thereby exercising a dominance over him that was destructive of the best interests of society itself. Hence, at that moment, he opposed the proposal of State education; urging that the whole length that the State should go was simply to *require* that every parent should see that his children got a good education, and, in case of need, should help with money for that end; but for the State to *provide* education would be tantamount to killing out originality: 'a general State education is a mere contrivance for moulding people to be exactly like one another.'

In view of the general situation, which he regarded as very pressing, Mill took up the cause of the individual and pleaded that he should have the utmost freedom for development—freedom of thought, of speech, and of action. No doubt, the individual is a member of society by birth and upbringing, and therefore, of necessity, a social being, and the question of the extent of the power of society over him is a most important, though a rather delicate, one; but he is also a centre of life and energy, with endowments special to himself, and this fact must be duly acknowledged and safeguarded. And what applies to the individual applies to any body of individuals, who co-operate of their own free choice for some end or purpose, and whose co-operation refers solely to such things as concern themselves jointly, and do not concern any persons but themselves.

Mill's defence of freedom of thought and of discussion (greatly lauded by his fellow-utilitarians, George Grote and others) is elaborate and telling. In his little book *On Liberty*, he writes with eloquence and enthusiasm, as well as with power. Various cogent reasons are given why opinions held by any person adverse to those generally recognized in the community should be tolerated.

One is that, in summarily suppressing an opinion, whether by legal penalty or by public obloquy, you may thereby be suppressing the truth; for commonly-accepted opinion is not necessarily true—it may be false. It is neither general assent nor long-continued recognition and acceptance that constitutes the ultimate reason for our adhering to a doctrine or upholding an institution: custom and tradition have often gone to the support of error and of wrong, and, if their tyranny is to be broken, it must be effected by the arguments and opposition of the individual. This applies to all departments of thought and of sentiment—to religion and morals, as much as to politics and social usages. The rights of the individual are indefeasible and inviolable.

Again, living interest in a truth is fanned by conflict and opposition, and thereby the truth itself is saved from becoming, as accepted opinion so frequently does become, a dead, useless dogma. It is only by the constant need of meeting the negative that the full force and meaning of the affirmative is understood, and the individual's own views clarified. When compelled to defend a position, one is driven to clear, full, and consistent thinking, and to deliberate consideration of the weak points, as well as the strong points, in one's position. Opposition and reasoned denial may for the

moment be disconcerting, but it is intellectually in-
vigorating and helpful: it clears the upholder's ideas and
strengthens his convictions, and gives him greater
energy to maintain his ground and greater confidence
in propagating his opinions.

Once more, truth, as a rule, is not the sole possession
of either side in a controversy, but is shared in by both
sides. This arises from the very nature of the case.
Truth is infinite, and has many diverse aspects; but its
diverse aspects are not contradictory of each other (as
disputants so often think) but complementary, although
a fresh aspect, especially if forcibly presented, is apt to
appear for the moment destructive and upsetting solely.
The correct view can be reached only by joining the
complementary aspects together and doing full justice
to each.

These are strong grounds for general toleration of
opinions and for unhampered freedom of discussion;
and, while they testify to Mill's insight, they disclose also
his wide sympathy and his own tolerant spirit. The last
of them, in especial, brings out a well-known trait of
his character. More than most thinkers, he respected
the views of his opponents and tried to discover the
modicum of truth that each contained and to give due
credit for it. At one time of his life, indeed, he carried
this tolerance and courtesy to an extreme and unnecessarily toned down or tacitly passed by some of his
own doctrines—so, at any rate, his friends thought and
he himself (in later life) believed.

When, now, we turn from individual freedom of
opinion and discussion to freedom to carry out one's
opinions into action—freedom of conduct—there is a

speciality to be noted. In acting, we are usually affecting others as well as ourselves, though it may be only indirectly; and, in 'self-protection' (Mill's own word), they may refuse to allow us to do what they regard as harmful to themselves or to society. Consequently, if the ground on which restraint is to be enforced is that of 'self-protection', the limitation of the individual's actions or conduct is obvious. 'He must not make himself a nuisance to other people. But if he refrains from molesting others in what concerns them, and merely acts according to his own inclination and judgement in things which concern himself, the same reasons which show that opinion should be free prove also that he should be allowed, without molestation, to carry his opinions into practice at his own cost.' The same holds of groups of individuals, who act in voluntary concert in things concerning only themselves. The crucial point, then, lies here—to determine precisely what are the actions in which a man may indulge that concern only himself. With regard to a man's opinions, Mill had laid it down that, if a man stand solitary (like Athanasius) against the world, mankind has no right to silence him; and the reason is because 'the peculiar evil of silencing the expression of an opinion is that it is robbing the human race; posterity as well as the existing generation; those who dissent from the opinion, still more than those who hold it. If the opinion is right, they are deprived of the opportunity of exchanging error for truth: if wrong, they lose, what is almost as great a benefit, the clearer perception and livelier impression of truth, produced by its collision with error.' Does not something similar apply to a man's actions, proceeding from his genuine opinions and inclinations? Are they to be disallowed simply because some of them are distasteful to people living at the moment, or are contrary to current social convention? Mere social dislike, or the practice of people in general, is no infallible proof of the real worth of conduct. And even if an action or course of action seem in itself morally reprehensible, has it no value of a deterrent nature? Would not the conflict that it sets up with accepted morals be itself to the advantage of morality? In answer to this, Mill enunciated the doctrine of the desirability for society and for the race of experimenting in modes of living, and contended that 'the worth of different modes of life should be proved practically, when any one thinks fit to try them', short of injury to others. This is what individuality means: the individual's own nature and its development, and not traditions or customs of the people, must be the determining factor. If this be not permitted, if a man is not to be allowed to develop his own character at his own risk, then 'there is wanting one of the principal ingredients of human happiness, and quite the chief ingredient of individual and social progress'. All this proceeds upon the principle that men are differently constituted by nature, and that what suits one man's circumstances may not suit those of another and that, if a man be not permitted to develop in his own way, he may not properly develop at all, and the world may lose thereby.

Where, however, a man's conduct impinges on the interests and rights of others, society may rightfully step in and repress him; and society must also insist that each person perform his duties, and discharge his obligations as a social being. Every person is both individual and social, and each aspect has its own right- ful claims, which must be respected : all that is demanded is that society shall not claim everything.

Mill's doctrine of the individual's liberty of conduct may be summarized under three heads: (1) The ad- vocacy of the due recognition of the place and import- ance of impulse and desire in man, as distinguished from intellect, though in close connexion with it—the supreme need of amply acknowledging the active and energetic side of the individual's nature. (2) Insistence on the view that spontaneity or individuality is a necessary ingredient in happiness or human welfare. (3) Revolt against the conventionalities of society that hinder, or seem to hinder, the development and expression of individuality—against the despotism of social custom. His own conduct not infrequently exemplified this revolt, and, in consequence, he suffered in the public estimation. In special, he defied social opinion in his relations with Mrs. Taylor, and parted with former friends; and he lost his seat in Parliament, at the General Election of 1868, in large measure through his practical support of Bradlaugh, whose pronounced free-thinking was offensive to many. On the other hand, his independence was sometimes rewarded, as when he was elected to Parliament by the people in 1865, not- withstanding that he deliberately gave public expression to opinions that he knew to be distasteful to them, thereby braving their displeasure, and refused to canvass his constituents in the usually accepted fashion.

The independence that he claimed for himself he demanded for every other candidate for a seat in

Parliament. The people's representative, he held, must not be a mere echo of the people, but a personal intelligent force, able to guide and to instruct; and even when he has to waive his own opinion on minor matters (when no fundamental principle of morality is involved) in order to secure victory on matters of greater importance, he is counselled to let his own opinion be publicly known all the same; 'insincere professions are the one cardinal sin in a representative government' (*Letters*, II, 67).

There is much in Mill's defence of freedom of conduct for the individual that is admirable, and his working out of the theme is very skilfully done. But there are points that lend themselves to criticism; two of which may be here mentioned. One is that, in his argument, he is apt to identify the individual's energy with 'genius' or 'originality', and to forget that energy may be mere eccentricity—not strength of character, but weakness—and, consequently, something needing to be repressed, rather than encouraged. 'The eccentric man', as Sir Leslie Stephen puts it, 'is a cross-grained piece of timber which cannot be worked into the State.' 'Eccentricity', says Professor MacCunn, 'is but the parody of individuality.' The other point is, that Mill does not sufficiently recognize that, although a man's desires and impulses are indispensable to the development of his nature, they are not a sure guide to the proper outlet for his activity. Unrestricted liberty to experiment may, indeed, produce 'variety' of human character; but character is estimated by its quality, not by variety. Even a 'genius' may profitably learn something from the experience of others, and thereby increase his happiness: at any rate, it may be doubted whether his own peculiar bent is not best cultivated under the opposition and restrictions of society.

II. REPRESENTATIVE GOVERNMENT.—Although a pronounced Radical, Mill was very much alive to the weaknesses and dangers of democracy, and tried hard to provide against them and to counteract them. In especial, he was greatly disturbed by the inadequate representation of minorities in Parliament and the eagerness of the majority to tyrannize over the minority, and, under the influence of sectional or class interests, to perpetrate injustice and one-sided legislation. Minorities have rights, he keenly felt, as well as majorities; and if the voice of minorities is not duly heard in the government of the country, democracy cannot be in a healthy or satisfactory condition.

The magnitude of the danger may be seen even at the present moment, when we contemplate parliamentary representation. To any one scanning the number of votes cast, on this side and on that, at a general election, it is obvious that the victorious party (whichever side wins) have a larger number of seats assigned to them than the votes of their supporters justify. The disparity is most striking, perhaps, when a section of the country is taken—say, Wales or Scotland. At the General Election of 1906, the votes of the Ministerialists in Wales were 217,462, and those of the Unionists 100,547, the proportion being roundly 2 to 1; and yet 30 Welsh Ministerialists were returned to Parliament and not a single Unionist! This result, involving the disfranchisement of the minority, is appalling. In like manner, Scotland, in the General Election of 1910, sent 61 Ministerialists to

Parliament, with a total of 372,313 votes; whereas the Unionists, with no less a total than 277,183 votes, had only 11 seats. The inadequacy and injustice of the present system is still further borne out by a reference to by-elections, more especially when the contest is a three-cornered one.

To meet a situation like this, and to secure that majorities and minorities shall have just the representation that each is entitled to, Mill supported the system of proportional representation, which he regarded as the necessary complement of democratic government. It is the system of the transferable vote, first proposed in Parliament by Thomas Hare, and strenuously advocated, at the present time, by the non-party organization known as 'The Proportional Representational Society'.

But, besides the tendency of democracy to tyrannize over minorities, there was another dangerous tendency that much exercised the thoughts of Mill. He felt intensely the need of wise, educated, and enlightened legislators—cultured men, who had made a special study of politics, who knew what legislation means and whose opinions would be independent and their actions above any sinister or selfish consideration. The dignity and the efficiency of the House of Commons were very dear to him. Yet he saw, on the part of the democracy, a disquieting disinclination to give its true value to culture and to entrust the work that needed training and skill to trained and skilled minds. This determined his attitude towards two further matters, of great political importance.

First of all, while prepared to grant the electoral vote to adults duly qualified, he keenly realized that the value of votes is not equal. Intelligence, education, and superior virtue count for more than ignorance, stupidity, and indifferent character. Hence, he advocated plurality of votes to the higher educated citizens. He even went the length of drawing out a scheme and grouping citizens (according to professions, training, &c.), so as to show what classes should and what should not be allowed a plurality of votes; the basis of the classification being mental culture and moral qualities. And all this was done in the interests of democracy, and with a view to save it from a great danger—the danger of levelling down. It is superior intellect and high character, in his view, that can best save the State and best promote the interests of the electors; and 'I should still contend', he says, 'for assigning plurality of votes to authenticated superiority of education, were it only to give the tone to public feeling, irrespective of any direct political consequences'. Giving the tone to public feeling, creating the right atmosphere, counted with him for very much.

While thus, however, formulating a scheme that would so far save the educated from the class-legislation of the uneducated, he must needs provide against the possibility of the educated practising class legislation on their own account. And so he brought forward his plurality scheme with an important qualification— 'that it be open to the poorest individual in the community to claim its privileges if he can prove that, in spite of all difficulties and obstacles, he is, in point of intelligence, entitled to them'. The proof was to be given by means of 'voluntary examinations'!

The other matter referred to was Payment of Members of Parliament. This Mill unhesitatingly opposed. His object again was to secure purity and efficiency of parliamentary action. The interest of democracy itself forced upon him consideration of possible or probable consequences, and made him antagonistic to a practice that had many obvious elements of evil inherent in it. At the same time, he maintained that unnecessary demands should not be made upon the material resources of the members, and so insisted that no expenses incurred at an election should be charged to the candidate himself.

Another burning question in Mill's day was that of the Ballot. Both Bentham and James Mill had been eager supporters of 'vote by ballot', and the article was part of the utilitarian creed. Not so J. S. Mill. He judged the ballot, as was his wont with political proposals, by the *spirit* of it, by the kind of moral atmosphere it would introduce; and he condemned it. His great argument was, that voting by ballot was wrong in principle; it countenanced the claim of the voter to the franchise as a *right*—something that he was at liberty to dispose of exactly as he himself chose, without regard to the interest of any other person; whereas possession of a vote indicates a *trust*, and not a mere *right*—something to be used under a deep sense of responsibility, not for personal advantage, but for the general good. Grant absolute secrecy to the voter in recording his vote, Mill reasoned, and you remove from him the constraining consciousness that others have an interest in how he discharges his electoral function, and you encourage him to give rein to his selfish desires. Consciousness of responsibility to no one but oneself is a precarious guarantee of right action. This high moral platform is undoubtedly impressive. But, on the other hand, it may be asked, What of bribery and intimidation of the voter, whose operation the ballot was designed to check or to render impossible? Mill answered that bribery and intimidation were steadily on the decrease, and that he had every confidence that, in a very short time, they would be practically non-existent. In this, most people will think, he was too optimistic. At any rate, the real crux of the position lies here—Is it a greater evil to exercise one's vote secretly, in accordance with one's wishes, even if regardless of the opinion of others, or, it may be, contrary to their interests, than to cast it under intimidation or coercion (and, therefore, contrary to one's convictions)? Later political judgement has determined that it is the lesser evil to run the risk of secrecy, believing that even this risk may be more theoretical than real.

The Suffrage was another burning question. Mill urged the extension of the franchise to adults, on the ground of its powers to cultivate the minds and sentiments of the masses. Bentham had laid it down that a citizen, in order to be qualified as a voter, should be able to read. Mill went farther and said—read, write, and count. He would grant the suffrage to all adults (women as well as men) of adequate age, conditioned by the fact that the voter is a taxpayer, not under legal disqualification—such as that of being participator in parish relief. The suffrage that he contemplated, therefore, was to be *universal* but *graduated*. The requirement that the voter should be a taxpayer necessitated the lowering of the amount of tax demanded, if the power of voting were to reach the poorer classes. Grote had a very definite proposal for this object, as we shall see farther on.

Mill does not discuss the question of the Monarchy, but he gives frankly his opinion of the House of Lords. While maintaining the general necessity of such a House, he ingeniously suggests, as a reform, a special use of the Second Chamber. As its superior strength lies in its legal ability, the framing of Bills to be brought before Parliament should, he thought, be entrusted to it. From the nature of its constitution, the House of Commons is little competent for the drafting of Bills, though it must reserve to itself the right of final revision.

These views show Mill as a wise and practical legislator, fully alive to the dangers of popular government and animated by a strong sense of justice and by an earnest desire to put morality in its right place in political legislation. His wisdom has had its effect; and even now appeal is constantly being made to Mill's teaching whenever progressive measures, affecting the character as well as the material interests of the nation, fall to be considered.

CHAPTER XI

GEORGE GROTE; JOHN AUSTIN; ALEXANDER BAIN

WE have now the substance of the teaching of the Utilitarians—social, political, economical and philosophical—during the period with which this little treatise is concerned. Beyond this limit, we should have had to trace further developments, as in Herbert Spencer and Sir Leslie Stephen, under the influence of the conception of Evolution; in a peculiarly interesting presentation of utilitarianism, in a setting of Neo-Hegelianism, by T. H. Green; and in still another exposition, from a different standpoint, conciliatory in its purpose, by Henry Sidgwick. But, even within the limit of our study, several other names fall to be noticed.

I. GEORGE GROTE.—One is George Grote (1794–1871), the brilliant historian of Greece, the erudite expounder of Plato and of Aristotle, and a politician who did good work in the interests of reform before the victory of the first Reform Bill in 1832. He was a Benthamite of a very pronounced type (having come under Bentham's personal influence at an early date), and his political views and leanings are manifest in all his chief writings. His learned works on Greek history and on Plato and Aristotle are the presentation of Greek thought and Greek political action very much as seen through the spectacles of the British philosophical Radical of the nineteenth century. He was a practical politician, as well as a political thinker, and his name is specially associated with his insistent advocacy in Parliament, during the years that he sat as member for the City of London (1832–41), of Vote by Ballot. On this subject, he separated himself from his fellow utilitarian, J. S. Mill, and stood forth as the loyal representative of orthodox utilitarian opinion. He did not share Mill's belief in the continuous decrease of corruption and intimidation at parliamentary elections. On the contrary, he took careful note of the riotous ebullitions of feeling that disgraced election after election as time went on, and made telling use of them as an argument in favour of the Ballot. His handling of the subject is authoritative and exhaustive. Two of his arguments may be specially mentioned. While meeting the objections on which Mill relied, and other objections besides, he laid the greatest stress on the fact that the system of open voting current at the time was practically the disfranchising of hundreds, even thousands, of the electors. For a great many voters, not daring to brave intimidation, refused to vote at all; and others, voting at the dictation of a superior, urged by self-interest or other motive, were simply throwing a multiplicity of votes into the hands of the superior, instead of the single vote, or the limited number of votes (if he were a plural voter) which by rights belonged to him. The result was that the House of Commons failed to possess the full confidence of the people, and the object of representative government was thereby defeated. The other argument is that by which he disposed of the objection against

secret voting on the ground that it affords a person who
has pledged himself to an individual to vote contrary to
his convictions an opportunity of breaking his word.
His position is that, in such a case, there must be lying
in one or other of the two alternatives—whether the
person votes against his convictions and fulfils his
private pledge, or votes in accordance with them and
thus fails to fulfil his private pledge. But voting contrary
to a man's convictions is a breach of a trust to the public,
whereas breaking his private pledge involves only the
individual to whom the pledge is given; and, although
both alternatives are deplorable, it is the lesser of the
two evils to be unfaithful to the individual than the
public—to break a private pledge than to violate a
public trust. This was incontrovertible on utilitarian
principles.

Grote's views on the Ballot are given in his pamphlet
on *Essentials of Parliamentary Reform*, published in 1831,
and in various striking speeches in Parliament delivered
by him between the years 1833 and 1839 (see his *Minor
Works*, 1873).

Besides ardently supporting secret voting, Grote was
an eager champion of the extended franchise. One
point is peculiar to him. In order to meet the difficulty
about the tax-paying qualification, which had exercised
the thoughts of J. S. Mill, he advocated provision for
'gradually lowering the suffrage at the end of some
fixed period, say five years, so as to introduce successively
new voters at the end of every five years and to render the
suffrage at the end of twenty or twenty-five years, nearly
coextensive with the community'. He thought that the
period of time here suggested would be sufficient to

educate the poorer voters, and thus 'obviate all ground for alarm on the part of the richer'.

Grote was a very ardent upholder of the experiential philosophy, and of the utilitarian ethics, and, without making any striking addition to the teaching of the school, succeeded in putting the doctrines in a cogent and attractive fashion. His vast learning, his clear thinking, and his pointed style served him well, as may be seen by a reference to his reviews and papers in his *Minor Works*, and to his *Fragments on Ethical Subjects* (1876).

II. JOHN AUSTIN.—Another distinguished utilitarian of the Benthamite stamp was John Austin, the jurist (1790–1859). He did special service to the cause by elaborating, from the side of jurisprudence, the philosophy of law—the doctrine of sovereign authority and sanctions as giving the meaning of law in ethics, as well as in jurisprudence, and marking it off from the physical-science conception of law, which is simply the generalized statement of observed facts. In the sphere of nature, the conception of law as commanding, and enforcing obedience, has no meaning: in ethics and in jurisprudence, it is the leading idea. Here a central authority, duly constituted and recognized, issues decrees, and is prepared to follow up its orders by punishment, if necessary. To Austin, everything in thinking depends on clear conceptions and accurately defined terms. Hence, like Bentham and James Mill, he elaborated the handling of leading words and phrases, and rejoiced in drawing fine distinctions. His own chief function as a jurist was, as described by himself, that of 'untying knots'. In pursuance of this, he insisted, among other things, on the necessity of according a wide signification to the great utilitarian term 'experience', including in it the testimony of history no less than the experience of the individual; thereby giving a distinct impulse to the pursuit of the historical method, which has come to be recognized as invaluable to investigations in all the branches of science—mental, political, sociological, ethical, and theological.

Specially important is his theory of Government. He was sufficiently appreciative of the fact that governments do not come into existence full-formed and mature, but have to grow, and grow through 'the perception of the utility of political government, or the preference of the bulk of the community of any government to anarchy'. There has been no social contract, then, such as had been conceived by some philosophers to be the original basis of political society, and no insistence on 'the rights of man', but simply 'perception of utility'.

Austin's name and fame is mostly that of a utilitarian jurist—completing the work of Bentham and of James Mill. He was not a professional politician, and had no burning zeal for democracy. Indeed, he was distinctly conservative, and opposed parliamentary reform in 1859. As a jurist, he is assigned a very high place among authorities, and his treatise on *The Province of Jurisprudence Determined* (1832) marks an epoch in the science.

III. ALEXANDER BAIN.—A third name of very great distinction is that of Alexander Bain (1818–1903). Bain was the intimate friend and valued counsellor both of J. S. Mill and of Grote, and worked along with them in elaborating the associationist and utilitarian philosophy. He was not, however, a politician in the same sense as they were—that is, he produced no writings on political subjects or on economics, though his philosophical Radicalism was robust. His reasoned views on the science of politics are to be found in the fifth book of his *Logic*. On the other hand, he has a very distinct and all-important place in the school as psychologist, ethicist, and educationist. In his handling of education, he adds practice to theory, and was the first to face the question of the supply of educative material, as distinct from philosophizing on education. His educational position was due in part to the fact that for twenty years (from 1860 to 1880) he occupied the Chair of Logic and English in the University of Aberdeen, his native city. The requirements of the students in the English Class led him to devote attention to English Composition. The result was that, through his inspiring prelections in the lecture-room and his many published works on *Rhetoric* and *English Grammar*, and his treatise on *Education as a Science*, he exercised an enormous influence in the North of Scotland and thence outwards on education and in educational circles. But his greatest reputation is in the spheres of psychology and ethics. If James Mill was the first formal psychologist of the utilitarian school, Bain was his most distinguished successor; and if J. S. Mill brought utilitarian ethics to a head, Bain gave it the scientific form that was needed.

The psychology of Bain, like that of the other utilitarians, was purely associationist; and 'experience' was its watchword. Moreover, he was under the spell of the physical and physiological science of his time, and was able to treat the mind in direct connexion with the body, especially with the brain and nervous system, in a way that had not been done—indeed, that had scarcely been possible—before; and to introduce the natural history method into the description of mental phenomena and processes. As, moreover, he had an exceptionally intimate knowledge of many of the sciences (physical, mathematical, and biological), he was able to illustrate his subject copiously and with rare felicity from the realm of science; thus making his psychology particularly striking and suggestive. This may be seen in his two great treatises, *The Senses and the Intellect* (first published in 1855) and *The Emotions and the Will* (1859). It is characteristic of him, in tracing the development of knowledge, to begin with sensation and accord the chief place among the senses to the muscular sense, and to lay stress on the native spontaneity of the bodily organism—a spontaneity that does not depend on external stimulus, but originates in the fullness of the nervous centres. It is the discharge of surplus energy, giving rise to random movements, which in turn produce comfort or discomfort, pleasure or pain. In these pleasurable and painful feelings, Will finds its origin, choosing the one and eschewing the other; for movements that bring pleasure are persisted in and sought after, those that bring pain are shrunk from or avoided. Instinct also has a special value for Bain: as being primordial to the human constitution, it is the basis of our acquisitions—mechanical (as in imitation) and intellectual. Given sensation, instinct, and the spontaneity of the system, and given retentiveness and the power of discrimination (agreement and difference) as native to the human mind, Bain undertakes to show how, through the working of association, in the two forms of contiguity and similarity, the whole of our mental possessions are obtained—our knowledge of external reality, and such complex conceptions as those of space and time. And not our intellectual possessions only, but our emotions and our volitions as well. Intellect, feeling, will—all come under the scope of association, and are entirely explicable on associationist principles. Presupposed is consciousness, or mental awakeness; and Bain devotes much care to the explication of consciousness, and to an exposition of the nature and working of relativity and its law. His handling of belief is noteworthy. Most psychologists treat of belief under intellect; Bain places it under will, inasmuch as the test of it is preparedness to act, although he was disposed, in later years, to modify his position. He starts by assuming primitive credulity in the individual —a tendency at first to believe everything: doubt or incredulity arises only when we are met by some check or hindrance. 'The number of repetitions counts for little in the process: we are as much convinced after ten as after fifty; we are more convinced by ten unbroken than by fifty for and one against.' Here, he parts company with James Mill. Belief to Mill was a case of inseparable association; to Bain it is uniform experience uncontradicted. Inseparable association is generated by the number of repetitions; belief is consequent on the absence of contradiction.

But, besides advancing the associationist psychology, Bain greatly buttressed the utilitarian ethics. He expounded the nature of pleasure and pain; formulating their mode of working in two well-known laws—self-conservation and stimulation. He carried the matter farther by offering a full and pointed analysis of happiness, which he defines as 'the surplus of pleasure over pain', and which is attained only when the susceptibilities of the mind are gratified to the utmost, and the susceptibilities to suffering spared to the utmost. This analysis is specially valuable to the utilitarian. It brings psychology at every point to bear upon the subject, and is based on a wide view of man's native susceptibilities (including the pleasure of malevolence, which Bain held to be originally inherent in human nature and markedly effective as a spur to action) and of the power of association; it shows a calm and soberminded estimate of the relative worth of the various 'goods' that men aim at; it lays adequate emphasis on the sympathetic side of man's nature as an important source of pleasure, but also, not infrequently, a source of pain and of suffering to the sympathetic person; and it explicitly recognizes the need of method in life for the individual, if he would secure the fullest possible happiness—a personal note, significant of Bain's own methodical nature and the systematic way in which he planned out his life and the persistence with which he carried his principle into daily practice.

But, further than this, Bain extricated utilitarian ethics from the embarrassment in which it found itself in regard to the relation of pleasure to disinterestedness. Bentham had been unsatisfactory here; and even J. S. Mill, while recognizing the existence of disinterestedness, had resolved it ultimately into a perception, or at any rate a feeling, of pleasure. Bain refuses to see in it anything of the nature of a disguised selfishness, but assigns it an effective and independent footing in human nature. 'So far as I am able to judge of our disinterested impulses', he says, 'they are wholly distinct from the attainment of pleasure and the avoidance of pain. They lead us, as I believe, to sacrifice pleasures, and incur pains, without any compensation. . . . It seems to me that we must face the seeming paradox—that there are, in the human mind, motives that pull against our happiness. It will not do to say that *because* we act so and so, *therefore* our greatest happiness lies in that course. This begs the very question in dispute. . . . This is the only view compatible with our habit of praising and rewarding acts of virtue. If a man were in as good a position under an act of great self-denial, as if he had not performed it, we might leave him unnoticed. If he has rather gained than lost by the transaction, he could dispense with any reward from us.'

Once more, utilitarianism owes to Bain a more satisfactory analysis of Conscience than had yet been given. The crucial point in conscience lies in the fact of obligation, with the feeling of authority accompanying. This Bain distinctly traces to the social character of man, and the bindingness of the commands of the State, enforced by punishment for disobedience. Will, sympathy, and the leading emotions are involved in the moral sense, but 'the finishing stroke' in it 'is due to Education and Authority'—a fact that constitutes the moral sentiment 'a distinct and peculiar phenomenon, different from all the other exercises of will, sympathy, and emotion, or any compounds of these'. His ethical teaching is to be found partly in *The Emotions and the Will*, and partly in the *Mental and Moral Science*.

From this brief account, it will be seen that Bain occupies a very definite and distinctive place in the utilitarian school. He strengthened and developed its doctrines, psychological and ethical, thereby supporting the political thinking of the philosophical Radicals; and he gave a distinctively scientific exposition of its principles, with application to practice—especially to education.

To the Utilitarian Radicals, thus passed in review, Britain owes an immense debt. Their views held sway for the greater part of the nineteenth century, and the result was awakened interest in psychological investigation and ethical discussion in the schools, and, in active politics, social reforms and beneficent legislation to an extent that had previously been unthought of. The benefit is being felt to-day. The spirit that animated them is still operating, and the lines on which social and political action is at present proceeding were largely laid down by them. Time has corrected much, has outgrown much, has discarded much; but the keen resentment of injustices that characterized the utilitarians, and their ever-active sympathy with the poor and the oppressed, and their enthusiasm for human welfare, are strikingly apparent still. Nor can the world afford to lose their insistence on the need for basing a political creed on a scientific knowledge and analysis of human nature, both as it shows itself in the workings of the mind and in the foundation of character ('ethology', as J. S. Mill called it), involving acquaintance with the human emotions as springs of action; or their devotion to economic investigations and their practical interest in jurisprudence. They carried forward their principles step by step, each great thinker adding something of permanent value. Progress was their watchword, and their enthusiasm for liberty and the public good supplied the driving power. That is what the present time inherits from them. They supplied to the world no complete philosophical system, but certain well-defined principles that have stood the test of results, and that still allow of indefinite beneficent application. In their eagerness for reform, they were often too critical and too merciless in their wish to destroy, without sufficiently appreciating the value of opinions, customs and institutions that were marked out for destruction. They had their defects, and their failures, but their face was ever towards the future. Their course was like the progress of opinions which J. S. Mill, in a sentence in his *Diary*, describes thus: 'The progress of opinion is like the advance of a person climbing a hill by a spiral path which winds round it, and by which he is as often on the wrong side of the hill as on the right side, but still is always getting higher up.'

BIBLIOGRAPHY

By Sir ERNEST BARKER, Litt.D., D.Litt., LL.D.

A.—THE WRITINGS OF THE UTILITARIANS

1. *The Works of Jeremy Bentham*, published by his executor, J. Bowring, 11 vols., 1838–43. Besides this general *corpus*, the following editions of particular works may be consulted:

   Comments on the Commentaries [of Blackstone], edited with an introduction by C. W. Everett, 1928.

   Fragment on Government, edited with an introduction by F. C. Montague, 1891.

   Theory of Legislation, edited with an introduction and notes by C. K. Ogden, 1931.

   Theory of Fictions, edited with an introduction by C. K. Ogden, 1932.

2. There is no collected edition of the works of James Mill. His most important writings on Utilitarianism are:

   Elements of Political Economy, 1821.

   Essays on Government, Jurisprudence, &c., 1828 (The *Essay on Government* was reprinted, with an introduction by Sir Ernest Barker, in 1937).

   Analysis of the Phenomena of the Human Mind, 1829.

   A Fragment on Mackintosh, 1835.

3. The most important writings of J. S. Mill are:

   A System of Logic, 1843.

   Principles of Political Economy, 1848.

   Essay on Liberty, 1859.

   Dissertations and Discussions, 1859–75.

   Considerations on Representative Government, 1861.

Utilitarianism, 1863.

An Examination of Sir W. Hamilton's Philosophy, 1865.

Inaugural Address at St. Andrews, 1867.

Autobiography (of particular importance), 1873, and the posthumous *Essays*, 1874.

Letters, edited, in 2 volumes, by H. S. R. Elliot, 1910.

A volume published by C. Douglas, entitled *The Ethics of John Stuart Mill*, 1897, contains important passages from Mill's writings, with introductory notes.

4. D. Ricardo's works were edited, with a notice of the life and writings of the author, by J. R. McCulloch, 1846.

5. John Austin's *Lectures on Jurisprudence* were edited by R. C. Campbell, 1869–85.

W. Jethro Brown's *The Austinian Theory of Law*, 1906, contains a reprint of some of the main lectures and of the *Essays on the Uses of Jurisprudence*, with good critical notes and a valuable excursus.

3. George Grote's *Minor Works* were published, with critical remarks, by Alexander Bain, 1873.

7. In regard to the later development of Utilitarianism, mention may be made of the writings of Alexander Bain (*The Emotions and the Will*, 1859, and the *Autobiography*, 1904), and of the writings of Henry Sidgwick (*The Methods of Ethics*, 1874: *The Principles of Political Economy*, 1883: *The Elements of Politics*, 1891).

B.—BOOKS ON THE UTILITARIANS

The best general works are the following:

1. Sir Leslie Stephen. *The English Utilitarians*, 3 vols., 1900 (the fullest account of the school, written with sympathy, understanding and scholarship. See also his *History of English Thought in the Eighteenth Century*, 2 vols., 2nd ed., 1881).

2. Elie Halévy. *The Growth of Philosophic Radicalism,* English Translation, 1928 (the most philosophic and fundamental study of 'the Westminster philosophy'), and *Histoire du Peuple Anglais au XIXᵉ Siècle,* Vol. I, 4th edition, 1930 (especially the third book, on *Les Croyances et la Culture*). Good bibliographies are attached to both of these books: the bibliography attached to the first, by C. W. Everett, deserves especial mention.

3. A. V. Dicey, *Law and Opinion in England,* 2nd ed., 1914 (contains a study, by a master of English Law, of the effects of Utilitarianism on English law and politics).

4. Alexander Bain. *James Mill, a Biography,* and *John Stuart Mill, a criticism with personal recollections,* both published in 1882, are both of peculiar intimacy and importance.

5. John Grote. *An Examination of the Utilitarian Philosophy,* 1870. (A discussion of Utilitarianism by the brother of George Grote, who was professor of moral philosophy at Cambridge.)

6. Coleman Phillipson. *Three Criminal Law Reformers,* 1923. (The reformers are Beccaria, Bentham and Romilly.)

7. A. Seth Pringle-Pattison. *The Philosophic Radicals,* 1907.

8. *The Dictionary of National Biography.* Articles on Bentham, the two Mills, Ricardo and Romilly.

Among writings which deal specifically with Bentham, the following may be noticed:

1. C. W. Everett. *The Education of Jeremy Bentham,* 1931 (New York).

2. C. K. Ogden. *Jeremy Bentham, 1832–2032* (Centenary Lecture), 1932.

3. Graham Wallas, articles on Bentham in the *Political Science Quarterly,* March 1923, and in the *Contemporary*

Review, March 1926: see also his *Life of Francis Place*, 2nd ed., 1918.

Among the contemporary writings which deal with the Utilitarians, the following may be noticed:

1. Lord Macaulay. Articles in the *Edinburgh Review*, March, June and October, 1829.

2. Sir James Mackintosh, *Dissertation on the Progress of Ethical Philosophy*, 1830.

1942

INDEX

Printed in Great Britain by
The Camelot Press Ltd., London and Southampton
3½ · 47